KT-508-918

Biology Today

A course for first examinations

Biology Today

A course for first examinations

M. D. Robson

*Head of science, Daneford Comprehensive School,
Bethnal Green, London
Vice-chairman, Biological Sciences Panel,
Metropolitan Regional Examinations Board*

A. G. Morgan

*Head of year, Hurlingham Comprehensive School,
Fulham, London
Formerly Deputy head of science, Daneford
Comprehensive School*

Macmillan Education

© M. D. Robson and A. G. Morgan 1980

All rights reserved. No part of this
publication may be reproduced or
transmitted, in any form or by any
means, without permission.

First published 1980

Reprinted 1981 (twice)

Published by
Macmillan Education Limited
Houndmills Basingstoke Hampshire RG21 2XS
and London
Associated companies in Delhi Dublin
Hong Kong Johannesburg Lagos Melbourne
New York Singapore and Tokyo

Printed in Hong Kong

British Library Cataloguing in Publication Data
Robson, M D
Biology today.
1. Biology
I. Title II. Morgan, A G
574   QH308.7
ISBN 0-333-22359-4

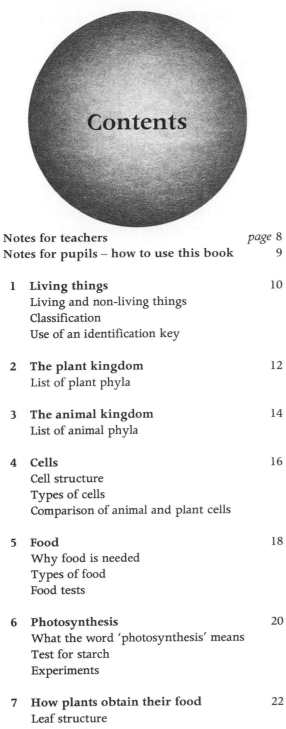

Contents

Notes for teachers

This book covers the contents of the mode I syllabuses of most CSE Examination Boards. Each of the seventy chapters is presented as a double-page unit with a careful synthesis of illustration and simple text. The chapters can be taught in any order, the sequence used by the authors is merely a suggestion which need not be followed. However, because of the natural inter-relation of the topics covered, an understanding of certain chapters may be reinforced and amplified by other chapters. Therefore, each chapter concludes with a short section headed 'Related Reading' which indicates other chapters in the book that will extend the pupils' understanding of the topic being studied.

Some pupils may wish to investigate particular topics to a greater depth than the scope of this book allows. Such pupils may refer to the 'Extra Reading' which is included at the back of the book. References have been restricted to those publications likely to be available in most school libraries. Throughout, the authors have made every effort to keep language simple and straightforward. However, some technical terms and 'difficult' words are inevitable. All such terms and words are printed in SMALL CAPITALS on the first occasion they appear in each and every chapter and are explained in the 'Glossary' provided at the end of the book.

Interspersed among the chapters are double-page sections of questions. These are designed to test understanding of the preceding material and to provide examination practice.

Although the authors fully recognise the fundamental role of practical work in any study of biology, it is felt that the organisation of practical sessions is best left to the individual teacher who can design his lessons according to the resources available. No detailed instructions are provided therefore.

The authors would like to thank Mr Nigel Baxter of the City of Leeds and Carnegie College for his helpful comments on the manuscript and Mr Norman Goodman, Head of English, Daneford School, who gave so much helpful guidance on language level. Although their assistance has been most helpful and warmly appreciated, the responsibility for any errors or shortcomings remains solely the authors'.

Some of the questions used in the book are drawn from past examination papers and the authors wish to acknowledge the assistance of the following Boards in allowing their questions to be reproduced:

Associated Lancashire Schools Examining Board (ALSEB)

East Anglian Examinations Board (EAEB)

East Midland Regional Examinations Board (EMREB)

Metropolitan Regional Examinations Board (Met. REB)

Middlesex Regional Examining Board (MREB)

North West Regional Examinations Board (NWREB)

South Western Examinations Board (SWEB)

Southern Regional Examinations Board (SREB)

Welsh Joint Education Committee (WJEC)

West Midlands Examinations Board (WMEB)

The West Yorkshire and Lindsey Regional Examining Board (WYLREB)

Yorkshire Regional Examinations Board (YREB)

The aforesaid abbreviations are used after questions in the book to denote the Board from whose papers they have been drawn.

The following boards have also allowed the use of multiple choice questions (numbers 1 to 7 in the Questions sections), which are not individually acknowledged: ALSEB, Met.REB, NWREB, SREB, and YREB.

Michael D. Robson
Ann G. Morgan

Notes for pupils— how to use this book

1 This book has seventy chapters and each chapter is two pages long.

2 A list of the chapter headings and a guide to their contents is on pages 5 to 7.

3 The chapters can be studied in any order. However, at the end of each chapter is a heading **'Related Reading'**. Here we list some other parts of the book which we think will widen your understanding of the chapter you have just read.

4 Difficult words are printed like this, IN SMALL CAPITALS. All words printed like this are explained in the **'Glossary'** on pages 171 to 174.

5 At nine places in the book there are two pages of questions. These questions can be used to test your understanding and to give you practice in various types of examination questions.

6 If you wish to find out more about a particular topic than this book allows, you will be helped by the **'Extra Reading'** list on pages 168 to 169.

1
Living things

Biology is the study of living things. All living things are called ORGANISMS and these organisms have certain CHARACTERISTICS in common.

It is easy to decide that organisms such as people, trees and insects are living. It is not so easy to decide whether very small things such as bacteria and viruses are alive. Biologists have listed seven characteristics of living things. If something is living it will have all seven of these characteristics at some stage in its life.

The seven characteristics of living things

1 Nutrition
All living things take in food regularly. This gives them energy and materials for growth.

2 Respiration
All living things release the energy from food by respiration. This usually involves taking in oxygen and giving out carbon dioxide.

3 Excretion
All living things must remove the waste materials produced during many of the processes involved in living. Many of these materials can be poisonous even in small amounts.

4 Growth
All living things get larger and usually more complicated.

5 Reproduction
All living things eventually die so if a SPECIES is to survive, it must produce new individuals. In some plants and animals only one parent is required – this is known as asexual reproduction. In others, two parents are needed – this is known as sexual reproduction.

6 Movement
All living things move under their own power. An animal can usually move its whole body. A plant may only move certain parts, such as the petals of a flower.

7 Irritability
All living things respond to changes in their surroundings. Such a response could be movement towards food or light or away from danger.

Non-living things, for example a motor car, may show some of these characteristics but only living things will show all of them.

Classification

All living things can be arranged into groups. This is called classification. The members of each group have certain common features.

The modern system of naming and classifying living organisms is based on the work of Karl Linnaeus, which appears in his book 'Systema Naturae', published in 1735.

Each living organism has two names: generic and specific. The generic name starts with a capital letter, and both names are normally printed in italics. An example is *Homo sapiens*, for which the common name is man.

The word species is used for a group of organisms which share a great many common features and which can normally interbreed successfully.

Similar species are collected into a genus.
Similar genera are collected into a family.
Similar families are collected into an order.
Similar orders are collected into a class.
Similar classes are collected into a phylum.
The phyla are collected into either the plant kingdom or the animal kingdom.
Plant phyla are sometimes called plant divisions.

Example: Man (common name)

Kingdom is Animal
Phylum is Chordates
Class is Mammals
Order is Primates
Family is Hominidae
Genus is Homo
Species is *Homo sapiens*

These 'specific names' are complicated and difficult to remember, but they are used in biology to avoid any confusion which could be caused by using 'common names'. There are many examples of a species having one common name in one part of the country and a different name in another part of the country, for example *Vipera berus* is sometimes called the viper and sometimes the adder.

Related reading

Chapter 69, Classification
Chapters 55–59, Using keys in ecology

Identification keys

KEYS can be used to identify an animal or plant. There are keys for each of the phyla of the animal kingdom. Try this example for an arthropod.

Key

①Legs absent .... go on to②
  Legs present .... go on to ③

②Very small with eight stumps................... ⑨
  Maggot-like... ⑫

③Numerous pairs of legs .... ④
  Three or four pairs of legs........................⑤

④One pair of legs per segment .... ⑦
  Two pairs of legs per segment...... Millipedes

⑤Four pairs of legs............................ Spiders
  Three pairs of legs ....⑥ Insects

⑥Wings present...⑧
  Wings absent .... ⑭

⑦Long animals, nine or more segments ...... ⑰
  Broad animals, seven pairs of legs. .Woodlice

⑧Two pairs of wings............................... ⑪
  One pair of wings .... True flies
                  (order Diptera)

Instructions

Look at the specimen and decide which of these two statements is correct. The specimen has legs, so 'Legs present' is correct. Go on to ③.

Statements ② are not needed for this specimen.

Look at the specimen and decide which of these two statements is correct. The specimen has three pairs of legs so go on to ⑤.

'Three pairs of legs' is correct for this specimen. Go on to ⑥.

For this specimen go on to ⑧.

'One pair of wings' is correct for this specimen.

The specimen therefore belongs to the order Diptera – the true flies. A key, just for Diptera, will be needed to find the species name of the fly.

2
The plant kingdom

Green plants make their own food, using sunlight, and so they are called PRODUCERS. They make this food by a process called PHOTO-SYNTHESIS. Plants are green because they contain a coloured substance called CHLOROPHYLL. Chlorophyll and sunlight are essential for photosynthesis.

The plant kingdom is divided into two groups, the non-flowering plants, such as seaweeds and mosses, and the flowering plants, such as roses and apple trees.

Non-flowering plants

Algae (phylum THALLOPHYTA)
A drop of pond water will contain hundreds of tiny ORGANISMS. Many of these will be single-celled green algae such as Chlamydomonas (see diagram 1).

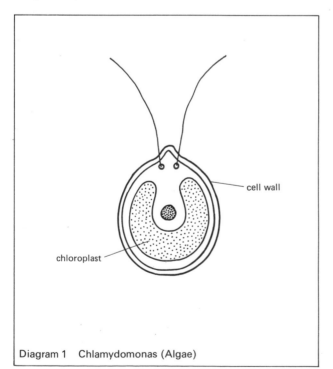

Diagram 1 Chlamydomonas (Algae)

Another algae, called Spirogyra, consists of cylindrical green cells joined end to end in a filament. Spirogyra is found in the slime on the surface of many freshwater ponds.

Marine (seawater) algae include most of the common seaweeds such as bladder wrack (*Fucus vesiculosus*). Many of these are brown algae. They still contain chlorophyll, but a brown-coloured substance is also present.

Fungi (phylum THALLOPHYTA)
Fungi do not contain chlorophyll and so cannot photosynthesise. For this reason, some biologists do not regard them as plants and put them in a kingdom of their own. Some fungi obtain their food from other living animals or plants and so are called PARASITES. Others live on dead or decaying material and are called SAPROPHYTES (see diagram 2).

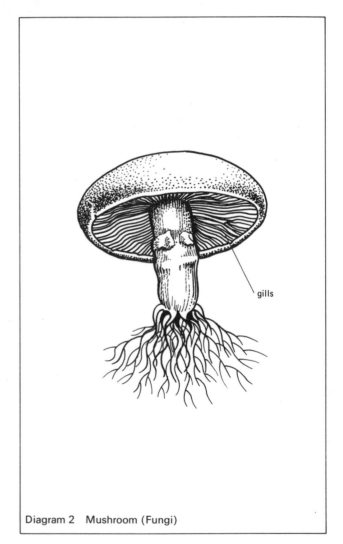

gills

Diagram 2 Mushroom (Fungi)

Lichens (phylum THALLOPHYTA)
Lichens can be found as a crust growing on rocks or old buildings in areas of clean air. They are a partnership of algae and fungi living together.

Liverworts and mosses (phylum BRYOPHYTA)
The liverworts have a flat, 'leaf-like' shape and can survive only in wet places. The mosses have a more complicated shape (see diagram 3) and they can survive drier conditions than can the liverworts.

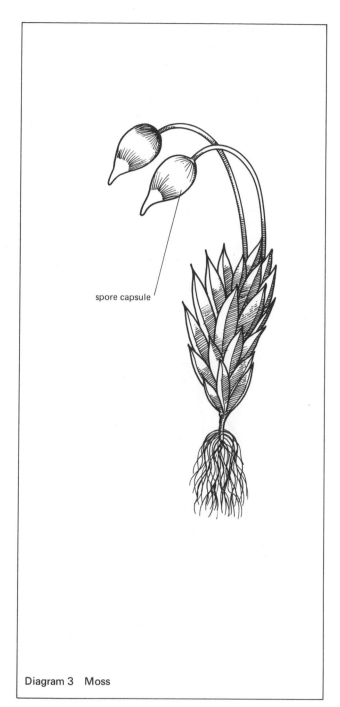

spore capsule

Diagram 3   Moss

cone

Diagram 4   Pine (Gymnosperm)

Ferns (phylum PTERIDOPHYTA)

Ferns have an underground stem with roots, and leaf-like STRUCTURES called fronds. Ferns reproduce by SPORES made in little brown lumps, called sori, on the underside of the fronds. Common examples of Pteridophyta are the brackens and the maiden-hair fern.

Gymnosperms (phylum SPERMATOPHYTA)

All spermatophytes produce seeds. They are divided into two groups, one of which is the non-flowering gymnosperms.

Gymnosperm means naked seed or seeds not enclosed in a fruit. This group includes the coniferous trees, such as pine, which reproduce by making cones to carry the seeds (see diagram 4).

Flowering plants

Angiosperms (phylum SPERMATOPHYTA)

The second group of seed-bearing plants contains the angiosperms or flowering plants.

The angiosperms are flowering plants that have seeds enclosed in some sort of FRUIT. They are divided into two sub-groups, MONOCOTYLEDONS and DICOTYLEDONS:

(i) Monocotyledons are plants such as daffodils, grasses and the cereals; they have narrow leaves with parallel veins. The word monocotyledon means 'one seed leaf'.

(ii) Dicotyledons have broad leaves with branching veins. The word dicotyledon means 'two seed leaves'. They can be grouped according to their size.

Herbaceous plants.	These are small plants such as daisy and buttercup.
Shrubs.	These are medium-sized plants such as privet and blackthorn.
Deciduous trees.	These are the larger plants such as oak, ash and beech.

Related reading

Chapter 6, Photosynthesis
Chapter 48, Reproduction in plants
Chapter 49, The flower

3
The animal kingdom

The animal kingdom is made up of ORGANISMS that depend on plants for food. Animals are called CONSUMERS because they eat the food that plants produce. Some animals such as lions get this plant food second-hand because they eat an animal which has eaten the plants.

The animal kingdom is divided into two groups. Animals with backbones, like horses and alligators, are called vertebrates and animals without backbones, like snails and butterflies, are called invertebrates. These two large groups are further divided into smaller groups called phyla.

Invertebrates (animals without backbones)

Phylum PROTOZOA

These are animals which are only one single cell. They are microscopic and live in water. Some move by lashing hair-like projections called CILIA or FLAGELLA, and others, for example Amoeba (see diagram 1), move by flowing.

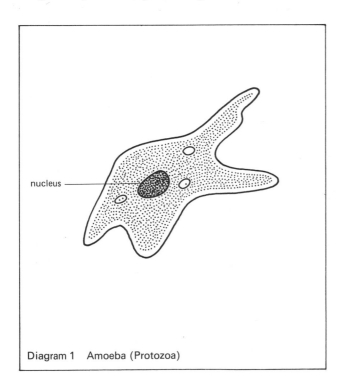

Diagram 1   Amoeba (Protozoa)

Phylum PORIFERA

These animals are known as the sponges and live in water.

Phylum COELENTERATA

These animals, such as the sea anemone and jelly fish (see diagram 2), have a single opening for food intake and waste output. There is a simple nervous system and all members of this phylum have tentacles with stinging cells.

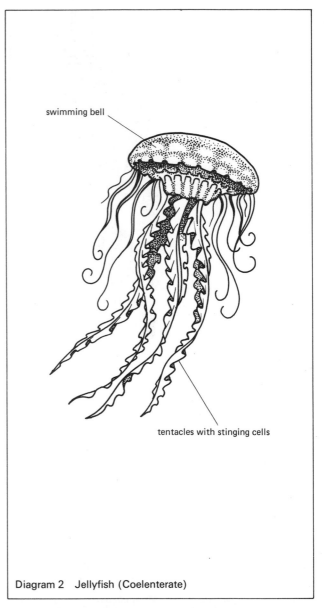

Diagram 2   Jellyfish (Coelenterate)

Phylum PLATYHELMINTHA (flatworms)

These simple animals have a head-end and a tail-end to the body. Some flatworms, such as planarians, live in fresh water and others, such as the liver fluke and the tapeworm, are PARASITES.

Phylum ANNELIDA (true worms)

The common earthworm, the lugworm and the leech are annelid worms. These animals have bodies which are divided into portions called segments. They have a simple gut, blood system and nervous system.

Phylum ARTHROPODA

The word arthropod means jointed legs. The arthropods are divided into four classes:

(i) **Myriapods.** These animals are found on land, usually in rotting plants. They have a large number of legs on a long body. Examples are centipedes and millipedes.

(ii) **Crustaceans.** The crabs, lobsters and shrimps belong to this class. They live mostly in the sea but the wood-louse is a crustacean that lives on land. Crustaceans have many legs (see diagram 3).

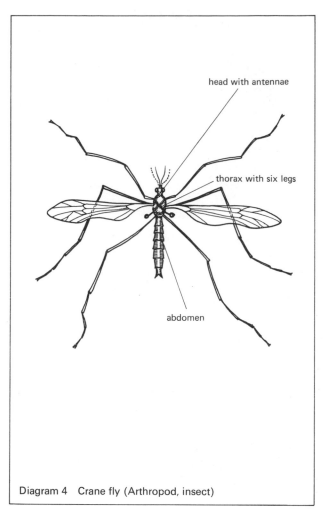

Diagram 4   Crane fly (Arthropod, insect)

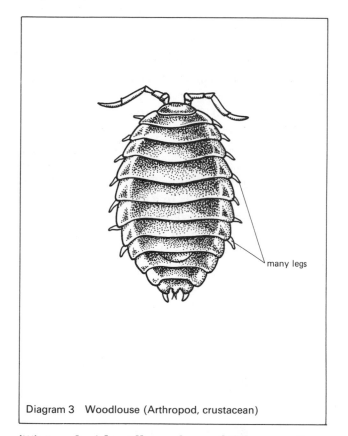

Diagram 3   Woodlouse (Arthropod, crustacean)

(iii) **Arachnids.** All members of this group have eight legs. The group includes the scorpions and the spiders.

(iv) **Insects.** All insects have six legs and three parts to their body (see diagram 4). There are over three-quarters of a million different SPECIES of insect.

Phylum MOLLUSCA

This phylum includes the snail and the octopus. The animal's body is not divided into segments and is often protected by a shell. Most have a muscular foot and a definite head. Other examples are the oyster, the winkle and the limpet.

Phylum ECHINODERMATA

These animals live in the sea. They are all animals with a circular body-plan, which means they have no definite left and right or head and tail. Examples are the star fish and the sea urchin.

Vertebrates (animals with a backbone)

Vertebrates have two pairs of limbs and a head with a brain. They are all members of the phylum CHORDATA but are divided into five different classes. Three of these have a variable body temperature:

(i) **Fish.** They breathe with gills and have a body covered with scales.

(ii) **Amphibians.** They can breathe on land and in water. Examples are frogs, toads and newts.

(iii) **Reptiles.** They live on land, have a dry skin with scales and include snakes, lizards and crocodiles.

The remaining two classes of the phylum CHORDATA have a constant body temperature and are sometimes called warm blooded:

(iv) **Birds.** They are covered with feathers and have a toothless horny beak.

(v) **Mammals.** They have body hair and feed their young on milk produced by special glands. Examples include rats, sheep, elephants, bats, whales and man.

Related reading

Chapter 8, How animals obtain their food
Chapter 43, Reproduction in animals

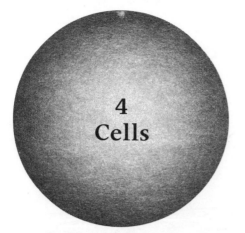

4
Cells

There are many different animals and plants. Each looks different in some way from the others, but there is something they all have in common. Every animal and plant is made up of cells which are often called the 'building blocks of life'. Cells are similar in the way they work and in the way they are made up.

Cells can be seen only under a microscope. Biologists have looked at many types of cell, from many different animals and plants. They have discovered that there are certain STRUCTURES which are always present.

The living world is divided into the animal and plant kingdoms and the reasons for this division can be found in the cells. The arrangement of structures is often different and plant cells have some extra structures not found in animal cells.

Cell structure (see diagram 1)

Each cell is like a factory which takes in raw materials, manufactures products and sends out what remains. Inside the cells are ORGANELLES, each of which does a special job.

The cell MEMBRANE separates the inside of the cell from outside fluids and other cells. This membrane is SEMI-PERMEABLE which means that only certain substances can pass through. In plants, a cell wall gives support similar to an animal's skeleton.

The VACUOLES are parts of the cell where there is mostly liquid. These are larger in plant than in animal cells. CYTOPLASM is the word used for the part of the cell where substances are built up and broken down. This process is called METABOLISM.

The cell must have energy to work. This energy is released in the MITOCHONDRIA by a process called respiration.

A very important part of the cell is the NUCLEUS. This contains a message from its parent cells, carrying instructions for how each cell is to behave. This message is in a code-form in chains of MOLECULES called CHROMOSOMES. The part of the chromosome which dictates a particular CHARACTERISTIC of the cell is called a GENE. The chromosomes can be seen inside the nucleus only when the cell is dividing.

Cells in the green parts of plants contain special organelles called CHLOROPLASTS. These

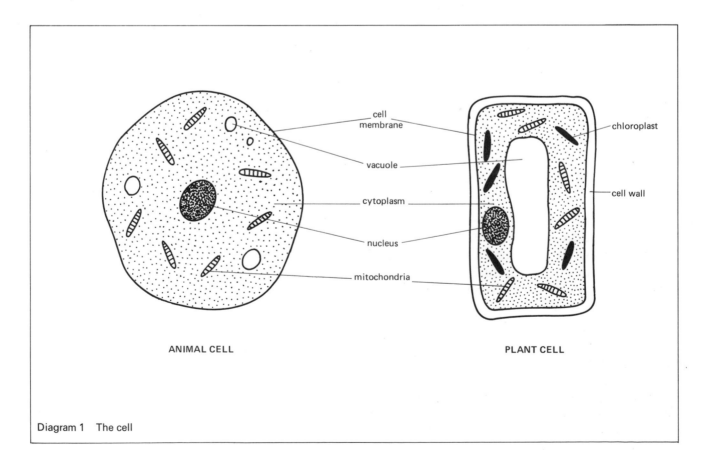

ANIMAL CELL   PLANT CELL

Diagram 1   The cell

can use the energy of sunlight to build up certain large molecules from small molecules.

Types of cells (see diagram 2)

There are many different types of cells. For example, root cells do a different job from leaf cells, and muscle cells do a different job from nerve cells. In very simple plants, like Spirogyra, and simple animals, like Amoeba, there is only one cell and it has to do every type of job. In many-celled ORGANISMS, like roses and elephants, cells can divide their labour. Cells doing the same job are sometimes grouped together and called TISSUE, for example muscle tissue. Different tissues working together can make up an ORGAN, for example a plant stem or an animal stomach.

Comparison of animal and plant cells

Plant cells	Animal Cells
1 Cell wall present	1 No cell wall
2 Cytoplasm around the edge of the cell	2 Cytoplasm usually fills the cell
3 Vacuole large	3 No vacuole, or very small vacuoles
4 Chloroplast present in green parts of plants	4 No chloroplasts

Related reading

Chapter 14, Cell membranes and cell walls
Chapter 15, Respiration in cells
Chapter 41, Growth and cells
Chapter 53, Genes and chromosomes

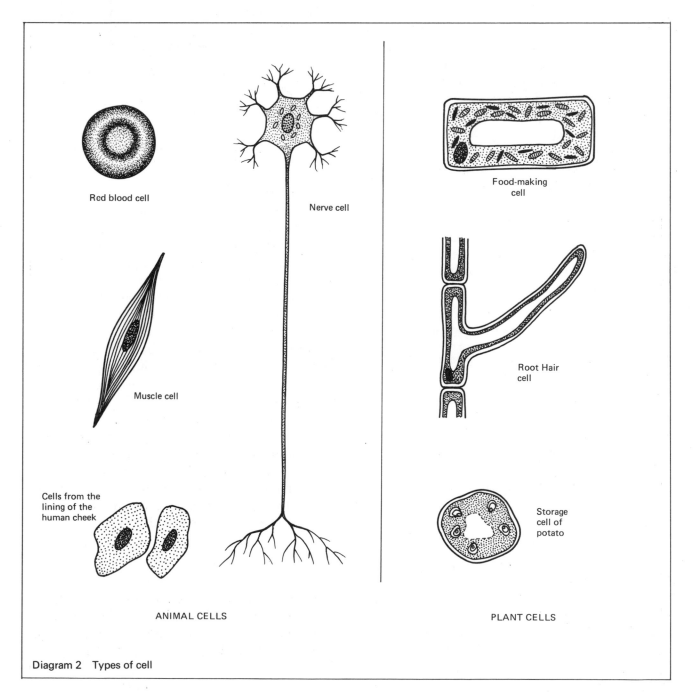

Red blood cell

Nerve cell

Food-making cell

Muscle cell

Root Hair cell

Cells from the lining of the human cheek

Storage cell of potato

ANIMAL CELLS

PLANT CELLS

Diagram 2   Types of cell

5
Food

When a box of matches is dropped and all the matches fall out, energy is needed to put them back in good order inside the box. Living things are carefully ordered with everything in a special place to do a special job. Energy is needed to keep things in order.

Some substances, like petrol, can be burned in an engine to give energy. This energy can do work, such as moving a car. People and horses can also move a car because they have energy. Energy is needed to do work.

Living organisms get energy from food, and not from fuels such as petrol or coal. Food as a fuel can be tested for the amount of energy it can give. Many fuels such as petrol and coal are tested for the energy they can give by burning them in plenty of oxygen and measuring the heat produced. Food energy is measured in this way in the laboratory using a calorimeter (see diagram 1 and table of results). The unit of energy used to be called the calorie, but now the name JOULE is used. (1 calorie = 4.2 joules)

Energy produced when 1 gram of food is burned

Foodstuff (1 gram burned)	Energy produced in joules
Bread	10 200
Cornflakes	15 300
Butter	33 400
Milk	2 800
Sardines	12 500
Pork	17 600
Chocolate	22 900
Lettuce	500
Runner beans	600

The results in the table show that some foods give a lot of energy and others give very little.

Living ORGANISMS need more than energy to keep them healthy and to grow. Food gives living organisms the raw materials they need to stay alive and reproduce themselves.

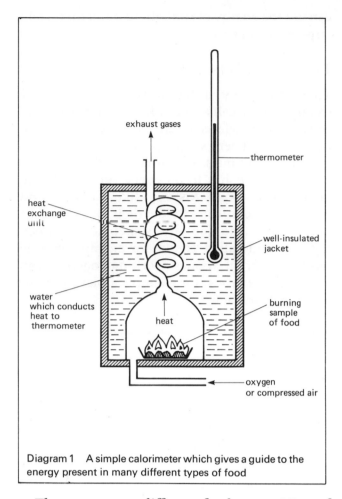

Diagram 1   A simple calorimeter which gives a guide to the energy present in many different types of food

There are many different food types. Most of the energy needs are satisfied by the carbohydrates and fats.

Carbohydrates are starches and sugars. These supply most of the energy requirement. A man chopping wood needs more of this type of food than does a man watching television.

Fats often form a store of energy which can be used at times when food is short.

Protein is the type of food that provides most of the new material for body building. A mother breast-feeding her baby and a fast-growing plant would use more protein than an old-age pensioner and an oak tree in winter.

As well as these three types of food, animals and plants need a diet which includes vitamins and MINERAL SALTS. These are needed in very small amounts compared with other food types, but no living organism is healthy without traces of these chemicals in its food.

Food tests

One meal may contain many types of food. Simple laboratory tests can give information about the main types present in a particular food-stuff. Chemicals, called reagents, are added to the food-stuff in a test-tube and a change may be seen.

Some food tests

Test for	Reagent	What is done	What is seen if the test is positive
Reducing sugars	Benedict's solution	Add and heat gently	Colour change blue→green→orange
Starch	Iodine solution	Just add	Colour change brown→blue-black
Protein	Millon's reagent	Add and heat very gently	Colour change colourless→pink-red
Vitamin C	D.C.P.I.P.	Just add	Colour change in liquid added blue→colourless
Fats	Ethanol and water	Just add	Cloudy solution

Vitamins and minerals important to man

Name	Sources (where to find this chemical)	Use in body	Diseases caused by lack of this chemical
Vitamin A	Dairy products Fish oils	Good eyesight Healthy skin	Poor night vision
Vitamin D	Butter Fish oils	Good bones and teeth	Rickets
Vitamin E	Wheatgerm Green vegetables	Affects reproduction	Possible sterility
Vitamin B_1	Egg yolks Marmite	Good nerves	Beri-beri
Vitamin B_2 (complex)	Liver Yeast Milk	Helps nerves of digestion and skin	Loss of weight Dermatitis
Vitamin C	Oranges and lemons Green vegetables	Healthy skin Helps resistance to disease	Scurvy
Calcium	Milk Green vegetables	Good bones and teeth	Soft bones and nails
Iron	Meat Spinach Potatoes	Good blood	Anaemia

Related reading

Chapter 8, How animals obtain food
Chapter 9, Balanced diet
Chapter 64, Food preservation

6
Photosynthesis

Animals and plants need food to remain alive. Most plants can build their own food from simple MOLECULES and are therefore called PRODUCERS.

Years ago it was believed that plants needed only water to grow. Today, biologists know that green plants and seaweeds use the energy of sunlight to produce food, in the form of glucose, from carbon dioxide and water. This process is called PHOTOSYNTHESIS. 'Synthesis' means to build up and 'photo' refers to using light. During photosynthesis, plants produce food and oxygen.

In most plants, photosynthesis takes place in the leaf. CHLOROPHYLL is the green colour in plants and is found in the CHLOROPLASTS of the plant cell. Chlorophyll must be present for photosynthesis to take place.

$$\text{carbon dioxide} + \text{water} \xrightarrow[\text{with chlorophyll}]{\text{in sunlight}} \text{glucose} + \text{oxygen}$$

The glucose produced during photosynthesis is stored in most plants as starch.

Test for starch in a leaf

1 The leaf is placed in boiling water to kill the cells and release the starch.
2 Chlorophyll is removed by carefully boiling the leaf in alcohol. This makes the leaf colourless but also brittle.
3 The leaf is replaced in boiling water to soften it.
4 The decoloured leaf is tested with iodine solution, giving a colour change of brown to blue-black if starch is present.

Experiments can be done to show what is essential for food to be made by a plant leaf. In the following experiments, photosynthesis is said to have taken place if starch is present in the leaf.

Destarching

Any healthy plant under normal conditions will have starch stored in the leaves. This starch must be removed before doing any experiments on starch production. Plants kept in the dark for about forty-eight hours will use up all the stored starch. No new starch can be made in the dark. The plant leaves are then said to be destarched.

An experiment to show that oxygen is a product of photosynthesis

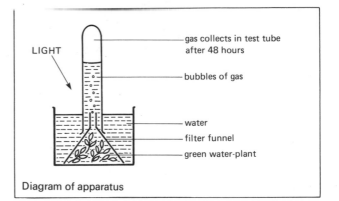

Diagram of apparatus

Water plants are used because of the problems of collecting a gas from land plants.

These water plants are in the light and have chlorophyll. There is carbon dioxide dissolved in the water.

The gas collected in the top of the test tube is tested for oxygen. A glowing splint is placed in this gas and relights. This test shows that oxygen is present in high concentration.
Result There is a great deal of oxygen present.
Conclusion Oxygen is one product of photo-synthesis.

An experiment to show that carbon dioxide is needed for photosynthesis

Carbon dioxide reaches the leaves of the plant from the air around the plant. Air contains small amounts of this gas.

A chemical called soda lime (a mixture of sodium hydroxide and calcium oxide) removes carbon dioxide from the air around the leaves.

The destarched plants are placed in bright sunlight for about four hours. Plant A has NO carbon dioxide; plant B is a CONTROL growing in ordinary air. Leaves from both plants are tested for starch.
Result Starch is NOT present in plant A. Starch is present in plant B.
Conclusion Starch is produced by the plant which has a normal amount of carbon dioxide around the leaves. A plant needs carbon dioxide for photosynthesis.

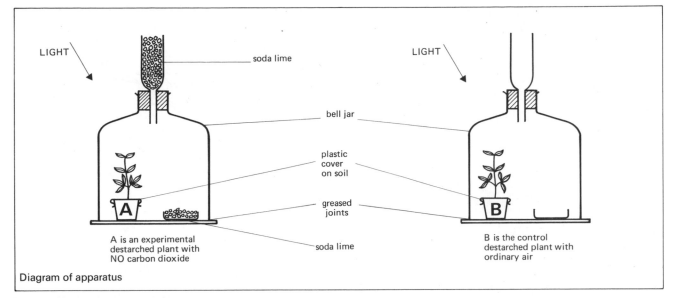

soda lime

bell jar

plastic cover on soil

greased joints

soda lime

A is an experimental destarched plant with NO carbon dioxide

B is the control destarched plant with ordinary air

Diagram of apparatus

An experiment to show that chlorophyll is needed for photosynthesis

The leaves of some plants are variegated, for example variegated privet. This means that they are not completely green, though they are quite normal and healthy leaves.

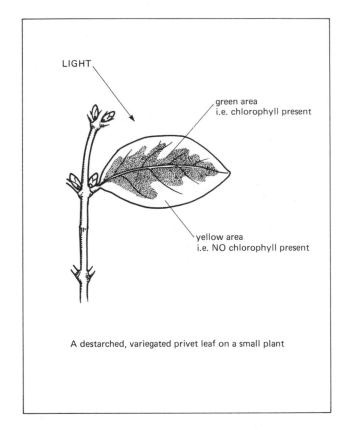

green area
i.e. chlorophyll present

yellow area
i.e. NO chlorophyll present

A destarched, variegated privet leaf on a small plant

A destarched variegated plant is kept in normal light for two days. The leaves are then tested for starch.

Result The colour change of iodine solution in the starch test shows that starch is present only in the green part of the leaf.

Conclusion Chlorophyll is needed for photosynthesis.

An experiment to show that light is needed for photosynthesis

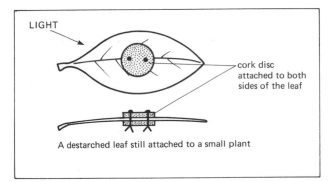

LIGHT

cork disc attached to both sides of the leaf

A destarched leaf still attached to a small plant

Cork discs are attached to both sides of several leaves on a destarched plant. This stops light from falling on the covered area of the leaf. The plant is kept in sunlight for several hours. The leaves are then tested for starch.

Result Starch is not present underneath the cork discs, but is present in the uncovered parts of the leaf.

Conclusion Light is necessary for photosynthesis.

Final conclusion

The experimental work demonstrates that starch and oxygen are produced by a plant when carbon dioxide, chlorophyll and sunlight are present.

Plants convert the energy of the sun into the chemical energy of starch. This can be used by all living organisms.

Related reading

Chapter 2, Definition of a plant
Chapter 7, Leaf structure
Chapter 21, Gaseous exchange
Chapter 68, Control experiments

7
How plants obtain their food

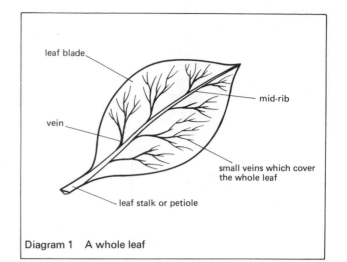

Diagram 1 A whole leaf

Plants can build the complex MOLECULES of starch from simple molecules of carbon dioxide and water using the energy of sunlight. This is called PHOTOSYNTHESIS and takes place in most plants in the leaf.

Leaf structure (see diagrams 1 and 2)

The epidermis is the outside layer of closely packed cells which usually contain no CHLORO-PLASTS. It protects the leaf from damage.

The palisade cells are large and contain many chloroplasts because most of the food building takes place in these cells.

The spongy layer contains many air spaces and the cells have fewer chloroplasts than the palisade cells.

The STOMATA are often more numerous on the lower side of a leaf. Stomata are openings to the atmosphere, allowing gases such as carbon dioxide, oxygen and water vapour to pass in and out of the leaf.

The guard cells on each side of the opening can swell or shrink and so open or close the stomata.

Plant needs

Plants need carbon dioxide, which they obtain from the atmosphere.
Plants need water and mineral salts, which they obtain from the soil.
Plants also need light.

Carbon dioxide

Living ORGANISMS produce about 500 million tonnes of carbon dioxide each year. In sunlight, palisade cells use up carbon dioxide to make food. Therefore the concentration of this gas in the outside air is stronger than the concentration inside the cell spaces. Carbon dioxide will DIFFUSE through the stomata and pass on into the palisade cells. MOLECULES are constantly moving from a place of high concentration, or strength, to a place of low concentration, or weakness, until the concentration of any particular kind of molecule is balanced in a whole system. This movement is called diffusion.

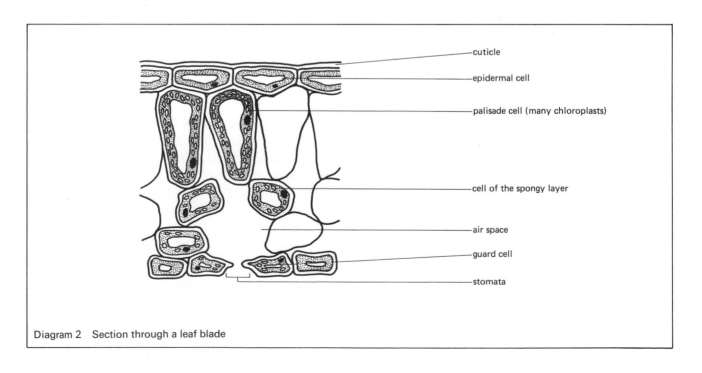

Diagram 2 Section through a leaf blade

Water

Most plants are about eighty per cent water. The air spaces inside the leaf are always SATURATED with water vapour. The air outside the leaf is seldom saturated, so water vapour diffuses out from the leaf. An average size birch tree loses about 370 litres (approximately 80 gallons) of water in a day in this way. This loss of water is called TRANSPIRATION. If a plant loses too much water it will WILT and the leaves will droop, which means that they catch less sunlight. To remain healthy, a plant must therefore replace the water lost by transpiration. Land plants have a root system which is in contact with soil water. The movement of soil water to the leaves, by DIFFUSION, is called the transpiration stream. The uptake of water by a plant can be measured using a POTOMETER (see diagram 3).

Conditions affecting transpiration

These are the same as the conditions affecting clothes drying on a line.

Humidity   A humid atmosphere is one in which there is a high concentration of water vapour. In humid conditions transpiration rates are low, and in dry conditions transpiration rates are high.

Temperature   Warm air takes up more water vapour than cold air. Therefore warm air gives higher rates of transpiration.

Air movements   Wind moves the air around the leaves and so any saturated air will be replaced by dryer air. Thus transpiration is greater in moving air than in still air.

The highest rate of transpiration takes place when it is hot, dry and windy.

Mineral salts

Many biologists believe that a plant uses energy to take in mineral salts from the soil. This is a complicated process called 'active uptake of ions'. Plants need these minerals as raw materials to build up the plant cells.

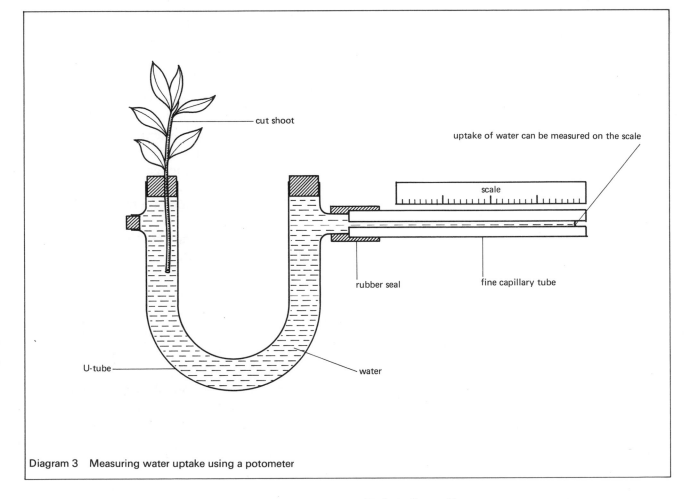

cut shoot

uptake of water can be measured on the scale

scale

rubber seal

fine capillary tube

U-tube

water

Diagram 3   Measuring water uptake using a potometer

Related reading

Chapter 6, Photosynthesis
Chapter 14, Diffusion
Chapter 21, Transport in plants
Chapter 25, How plants regulate water

Questions A

The answers to questions 1 to 7 are shown by one of the letters A, B, C, D or E.

1 Which of the following pairs contains only invertebrates?

A Amphibian, annelid
B Bird, insect
C Arthropod, reptile
D Insect, mollusc
E Fish, mammal

2 A plant without chlorophyll could be:

A an algae.
B a fungus.
C a gymnosperm.
D a liverwort.
E a coelenterate.

3 Which of the following animals is NOT an arthropod?

A Cockroach
B Spider
C Ant
D Lizard
E Crab

4 Which of the following is found in ALL living plant AND animal cells?

A Cellulose
B Chloroplasts
C Starch
D Cytoplasm
E Vacuole

5 Scurvy may be prevented by including in the diet sufficient:

A liver.
B cheese.
C fish.
D oranges.
E bread.

6 For photosynthesis, the gas taken from the air by the leaves of a green plant is:

A hydrogen.
B carbon dioxide.
C oxygen.
D water vapour.
E nitrogen.

7 Water loss from leaves occurs most rapidly when the weather conditions are:

A hot, humid and still.
B cold, wet and windy.
C hot, dry and windy.
D cold, dry and windy.
E mild, dry and still.

8 Copy out and then complete the following sentences:
(i)   Animal cells are bound by a plasma membrane whereas plant cells are bound by an additional structure called a ....
(ii)   The gas produced by plants in photosynthesis is ....
(iii)   The food material which supplies the growth and repair needs of the body is called ....
(iv)   The loss of water vapour through the stomata of leaves is called.............................
(v)   The vitamin C content of different foods can be estimated by the use of.........................
(vi)   The name of the gas that rekindles a glowing splint is ....
(vii)   One characteristic of a mammal is..........
....
(viii)   The crab is a member of the class of animals called....................................

9 Name the odd one out biologically in each group of words, and give your reasons:
(i)   Robin, hawk, bat, eagle, blackbird
(ii) Rose bush, grass, oak tree, privet, fern
(iii) Earthworm, goldfish, frog, snake, crocodile
(iv) Crab, beetle, starfish, shrimp, spider
(v) Seaweed, mushroom, moss, algae, liverwort

10 Write about six lines on FIVE of the following:
(i)   The leaf   (vi)   Chromosomes
(ii) Invertebrates   (vii) Transpiration
(iii) Energy   (viii) Food tests
(iv) Plants as   (ix)   Flowering
     producers   plants
(v) Mammals   (x)   Carbohydrates

11 Write an account, about twenty-five lines long, on ONE of the following. Diagrams may improve the answer.
(i) Water in plants
(ii) The importance of photosynthesis to animals
(iii) Classification
(iv) Cells
(v) Vitamins and minerals

12 Look at each of the drawings, (i) to (v), of the larvae of common freshwater insects (not drawn to scale):

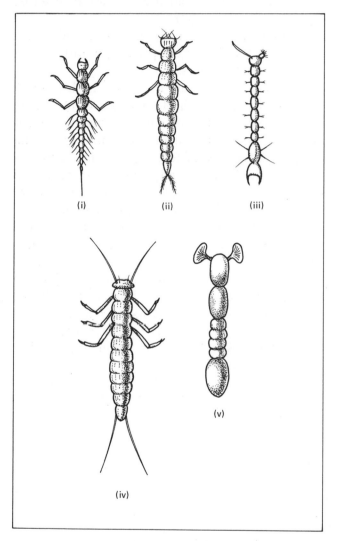

Identify each one in turn by using the key.
① No legs ....②
   Three pairs of legs present ....③

② Bristles on all segments .... Culex
   Bristles not present on all
   segments .... Simulium

③ One tail filament .... Sialis
   Two tail filaments present...................... ④

④ Long filaments on head........................ Perla
   Short filaments on head .... Dytiscus
                                          (Met. REB)

13 Draw diagrams of a typical animal cell and a typical plant cell. Label the following on your diagrams:

A nucleus   D cell membrane
B chloroplasts   E cytoplasm
C cell wall   F vacuole

14 The diagrams below are of a root hair and a transverse section of a leaf:

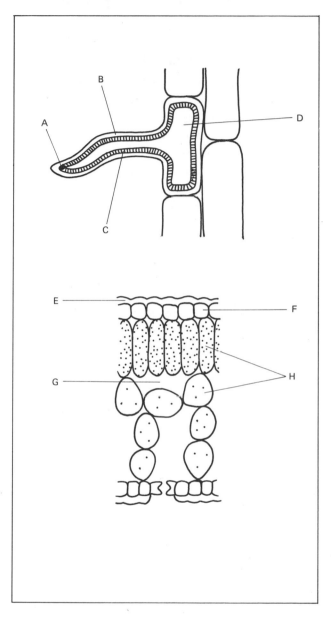

Name the Parts A – H.
                                          (Met. REB)

15 (i) Draw the apparatus of an experiment set up to show that carbon dioxide is needed for photosynthesis in the leaves of a potted plant.
(ii) What chemicals would you use to test a leaf for the presence of starch? What colour change would you look for?
                                          (Met. REB)

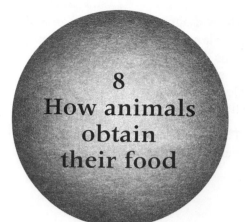

8
How animals obtain their food

Animals need food and water to remain healthy and to survive. Food is needed for energy and for body building. Water is an essential part of the living cell. The water lost through the removal of waste and through respiration must be replaced. Energy-giving foods are the carbohydrates and the fats. Body-building foods are the proteins. Animals also need MINERALS such as calcium and iron, and vitamins such as vitamin C and vitamin A. Proteins, minerals and vitamins make the chemicals which are part of the living cells of the body.

Plants can make their own food from simple chemicals, but animals must eat plants to get their food. They do this either directly, such as rabbits eating grass, or indirectly, like eagles eating rabbits which have eaten grass. For this reason, plants are called PRODUCERS and animals are called CONSUMERS. The food is transferred along what is called a food chain.

Examples:

grass seeds ──────→ woodmice ──────→ owls
seaweeds ──────→ periwinkles ──────→ seagulls

These chains can also be shown as a pyramid of numbers or MASS (see diagram 1). The length of the box is used to show either the number or the weight of each kind of ORGANISM.

The problem with using food chains and pyramids of numbers is that a living situation is not that simple. Many types of animals eat grass and many types of animals eat rabbits. The food relationships in a woodland, the sea or a pond can be very complicated and are shown better by a food web (see diagram 2).

One fact always remains the same and that is that plants are always the producers and animals always the consumers.

Not all animals obtain their food by killing. In the soil, in the sea and in fresh water, small animals like mussels and earthworms feed on the dead and decaying remains of plants and other animals. This type of feeding is called DETRITUS feeding and forms an important part of a food web. Plants such as fungi and bacteria which live on dead material are called SAPROPHYTES.

There are certain animals which live on, and sometimes inside, larger animals and plants. They are called PARASITES, and obtain their food from the efforts of the larger animal called the HOST. Such a parasite is the human head louse, which sucks blood for its food. There are also certain parasitic plants, such as the mistletoe and the dodder, which live on other plants, but parasitic plants are rare.

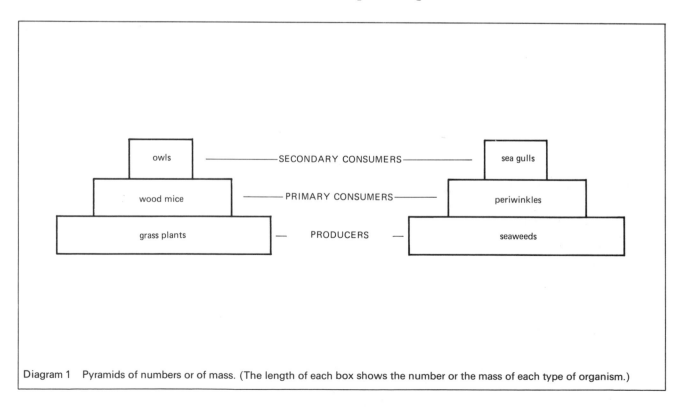

Diagram 1   Pyramids of numbers or of mass. (The length of each box shows the number or the mass of each type of organism.)

It can be seen that animals obtain food in a way which, in the end, always depends on plants. The pathway that food takes from plants, through a number of animals, back to the soil and the atmosphere is a cycle of materials such as carbon and nitrogen. However energy is always used up. This energy must be continually replaced by the sun through the process of PHOTOSYNTHESIS in plants.

Related reading

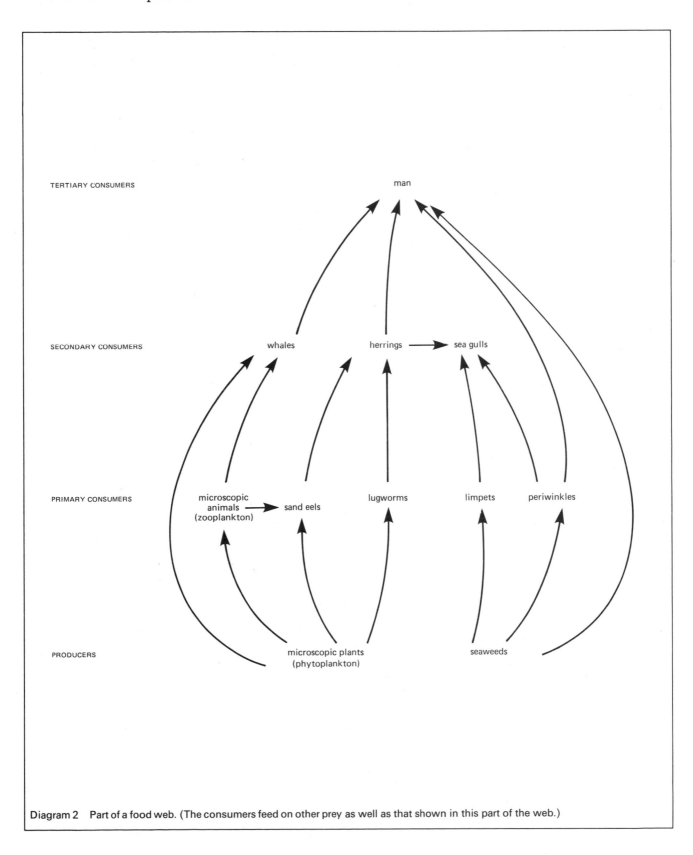

Diagram 2   Part of a food web. (The consumers feed on other prey as well as that shown in this part of the web.)

9
A balanced diet

A glass of milk contains carbohydrates, fats, proteins, calcium, vitamins A, B, C, and water. The human body needs these substances and uses the energy present in them to do work.

An egg will give most of these substances except carbohydrates and vitamin C. Sugar will give energy alone. So a selection of different foods is needed to keep the body healthy. This selection is called a balanced diet.

Carbohydrates and fats

The energy value of food is measured in JOULES (4.2 joules = 1 calorie). A thousand joules is called a KILOJOULE (kJ).

An adult man uses about 8000 kJ a day just to stay alive because the heart needs energy to pump the blood and the chest muscles need energy for their part in breathing. If he spends all day sitting down he will need only another 500 to 1000 kJ, that is 9000 kJ in total. If the same man spends all the next day working in the garden, his energy needs will be an extra 7000 kJ, that is 15 000 kJ in total.

A small child needs less energy than an adult. For example, a three-year-old boy may need about 5000 kJ a day.

Foods with a high fat or carbohydrate content are high in energy. If a man changes from a hard manual job, such as being a coalminer, to an office job then he should eat less high energy foods to avoid becoming overweight.

A diet which has sufficient amounts of carbohydrate and fat for energy, but not enough protein, vitamins or MINERAL SALTS can lead to illness and death.

Proteins

Proteins are used for growth and repair in the body. For every KILOGRAM of body weight, the body needs 1 GRAM of protein each day. A man who weighs 70 kilograms therefore needs about 70 grams of protein each day. A growing child needs 3 or 4 grams of protein a day per kilogram of body weight. A pregnant woman or a breast-feeding mother needs extra protein to feed the baby.

Vitamins

Vitamins are complex chemicals which have no energy value but are essential in small quantities to keep the body healthy. If a diet lacks one or more vitamins it causes disease. The disease can usually be cured by including the necessary vitamins in the diet. There is a table of vitamins and related diseases in chapter 5.

Minerals

The body needs many different MINERALS such as magnesium, potassium, sodium and phosphates. These are present in most diets, but calcium, iron and iodine can often be short.

Roughage

Roughage is also important in a balanced diet. It consists of the undigestible materials such as plant cell walls, which add bulk to the materials in the digestive system. This helps prevent constipation and its related problems.

Water

Man can live for only a few days without water. Water makes up about seventy per cent by weight of the body. It is essential for all chemical reactions and transport in the body. Man loses between two and three litres of water a day and this must be replaced from the foods he eats and the liquids he drinks.

A balanced diet must contain correct amounts of:

Carbohydrate
Fat
Protein
Vitamins
Minerals
Roughage
Water

The table (opposite) shows the amounts of these in some common foodstuffs.

Related reading

Chapter 5, Food
Chapter 66, Personal hygiene and balanced diet

Energy values of a meal
Total energy value = 5230 kJ

100 g peas (200 kJ)
200 g milk (560 kJ)
150 g canned peaches (400 kJ)
50 g cream (450 kJ)
Butter
180 g chips (1800 kJ)
250 g fried cod (1450 kJ)
20 g brown bread (200 kJ)
5 g butter (170 kJ)

10 grams of food	Energy (kilo-joules)	Carbo-hydrate (grams)	Fat (grams)	Protein (grams)	Vitamin A (micro-grams)	Vitamin B (micro-grams)	Vitamin C (micro-grams)	Vitamin D (micro-grams)	Calcium (milli-grams)	Iron (milli-grams)
Milk (whole)	28	0.5	0.4	0.3	0.8	0.4	21	0	12.0	0
Butter	334	0	8.5	0.04	29	0	0	0.01	1.4	0
Groundnuts	252	0.8	4.8	2.8	0	2	0	0	6.0	0.2
Beef	134	0	2.8	1.5	0	0.7	0	0	1.1	0.4
Cheese	177	0	3.5	2.5	8.4	0.4	0	0	81	0.07
Liver	61	0	0.8	1.7	120	3	315	0.01	0.7	1.4
Herring	99	0	1.8	1.7	0.8	0.4	0	0.2	10.2	0.1
Eggs	67	0	1.2	1.2	6.0	1.5	0	0	5.6	0.2
Bread (wholemeal)	97	4.7	0.2	0.8	0	2.1	0	0	2.5	0.28
Rice	150	8.6	0.1	0.6	0	0.7	0	0	0.4	0.04
Potato	37	2.1	0	0.2	0	1.0	210	0	0.7	0.07
Orange	15	0.8	0	0.07	0.4	1.0	492	0	4.2	0.04
Sugar (white)	165	10.0	0	0	0	0	0	0	0	0

1 gram = 1000 milligrams
1 milligram = 1000 micrograms

Table of food values

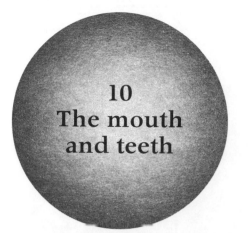

10
The mouth and teeth

In the human mouth, food is chewed and mixed with SALIVA. The saliva acts as a LUBRICANT, and also helps the food to stick together into a ball, called a BOLUS. This is pushed by the tongue to the back of the mouth and is swallowed.

Saliva contains an ENZYME called amylase, which starts the breakdown of food. Saliva acts on starch and converts it to maltose, which is a simple sugar. Chewing action by the teeth and tongue reduces the food to a suitable size for swallowing and mixes it with the amylase.

Teeth

There are four main types of teeth: incisors for cutting; canines for tearing; and premolars and molars for grinding. The arrangement of these teeth varies according to the diet of the animal. The number and arrangement of teeth is called the DENTITION of an animal. A sheep has a pattern suitable for nibbling short grass, but a dog has a different pattern suitable for tearing meat.

Some vertebrates have many teeth which are constantly replaced, but most mammals have a small number of teeth and only have two sets for a lifetime. The first set, or milk teeth, has no molars and is later replaced by the adult, permanent teeth.

Dental Formula

A dental formula shows the number of teeth of each kind in half the upper jaw and half the lower jaw. Only half the jaw is recorded as the left side is the same as the right side in all mammals.

Examples:
i = incisor, c = canine, pm = premolar, m = molar

Man

$$i\frac{2}{2} \; c\frac{1}{1} \; pm\frac{2}{2} \; m\frac{3}{3} = \frac{8 \text{ teeth in } \frac{1}{2} \text{ upper jaw}}{8 \text{ teeth in } \frac{1}{2} \text{ lower jaw}}$$
$$= 32 \text{ teeth in mouth (diagram 2)}$$

Pig

$$i\frac{3}{3} \; c\frac{1}{1} \; pm\frac{4}{4} \; m\frac{3}{3} = \frac{11 \times 2}{11 \times 2} = 44 \text{ teeth}$$

Dog (diagram 3)

$$i\frac{3}{3} \; c\frac{1}{1} \; pm\frac{4}{4} \; m\frac{2}{3} = \frac{10 \times 2}{11 \times 2} = 42 \text{ teeth}$$

Sheep (diagram 4)

$$i\frac{0}{3} \; c\frac{0}{1} \; pm\frac{3}{3} \; m\frac{3}{3} = \frac{6 \times 2}{10 \times 2} = 32 \text{ teeth}$$

Rabbit

$$i\frac{2}{1} \; c\frac{0}{0} \; pm\frac{3}{2} \; m\frac{3}{3} = \frac{8 \times 2}{6 \times 2} = 28 \text{ teeth}$$

The dog (see diagram 3) is a meat eater, or CARNIVORE. Its incisor teeth are pointed, and the canine teeth are large. This is a good arrangement for tearing meat from bones.

The sheep (see diagram 4) is a plant eater, or HERBIVORE. It has a hard pad on its upper jaw against which the lower incisors trap grass. There is a gap between the incisors and premolars called the diastema.

In many herbivores the teeth grow throughout the animal's life. This is because they have a persistent pulp which means the root remains widely open, allowing a good supply of food and oxygen to the tooth.

Man and pigs eat both plant and animal material, and are called OMNIVORES. Their teeth stop growing as the root becomes narrow.

Tooth structure (see diagram 1)

Enamel
This is the hardest substance made by animals. It is non-living and contains calcium salts. It covers the crown and neck of the tooth.

Dentine
This is fairly hard, but it cannot stand up to wear. Running through it are strands of CYTOPLASM from cells in the pulp cavity.

Pulp cavity
This is the living centre of the tooth containing blood vessels and nerves.

Root
The root is connected to the jaw socket by tough fibres which allow the tooth to move slightly. They act as 'shock absorbers' during chewing and biting. The fibres are attached to the dentine of the root by cement.

Tooth decay

The main cause of tooth decay is a sugary substance left on the teeth, which is converted into acid by bacteria. This acid attacks the enamel and allows the bacteria to enter the softer dentine. Cleaning the teeth regularly and eating less sugary foods can help to avoid tooth decay.

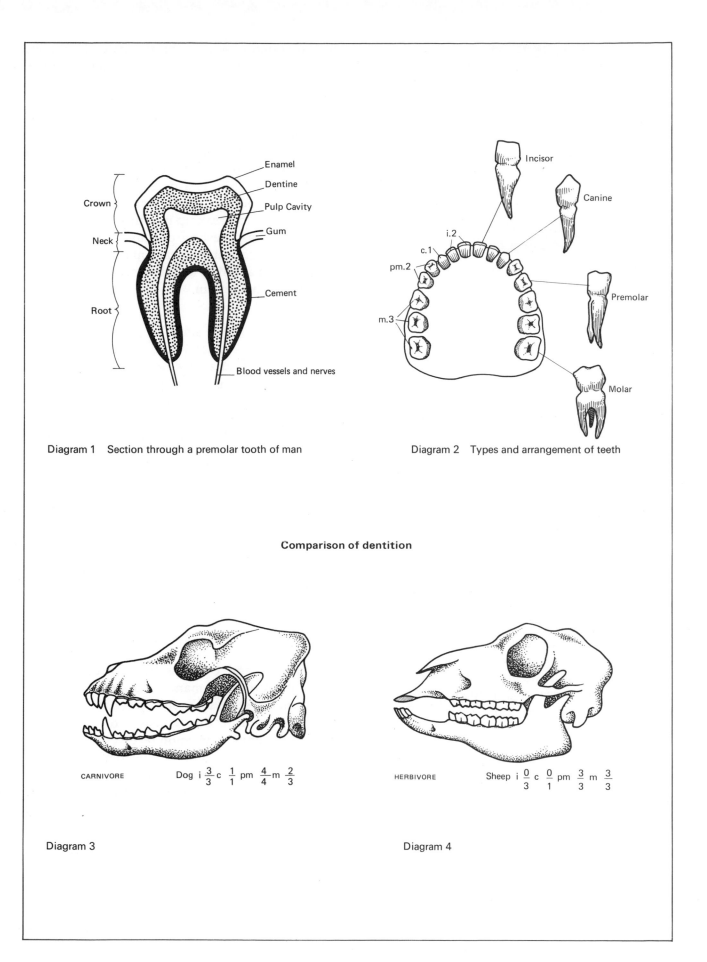

Diagram 1   Section through a premolar tooth of man

Diagram 2   Types and arrangement of teeth

Comparison of dentition

CARNIVORE   Dog   i $\frac{3}{3}$ c $\frac{1}{1}$ pm $\frac{4}{4}$ m $\frac{2}{3}$

HERBIVORE   Sheep   i $\frac{0}{3}$ c $\frac{0}{1}$ pm $\frac{3}{3}$ m $\frac{3}{3}$

Diagram 3

Diagram 4

Related reading

Chapter 12, Enzymes
Chapter 32, The tongue

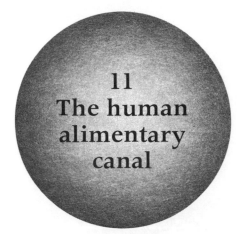

11
The human alimentary canal

The process of digestion takes place in the ALIMENTARY CANAL, which begins at the mouth and ends at the anus (see diagram 1).

Swallowing

The food BOLUS is pushed to the back of the mouth and swallowed. A flap of CARTILAGE, called the epiglottis, prevents the bolus going down the wrong way into the trachea (windpipe). A squeezing and relaxing action of the circular muscles in the oesophagus (gullet) produces a wave-like motion which forces the bolus down the oesophagus. This wave-like motion which moves the food through the alimentary canal is called PERISTALSIS (see diagram 2).

Stomach

Peristalsis carries the food into a muscular bag called the stomach. The walls of the stomach can expand allowing it to hold quite large amounts of food. The food is held in the stomach because the pyloric sphincter at the outlet of the stomach is closed. A SPHINCTER is a ring-shaped muscle band which can contract and stop the flow of material.

In the stomach the food is mixed with gastric juice secreted by glands in the stomach walls. Gastric juice contains hydrochloric acid, the ENZYMES pepsin and rennin which need acid conditions, and MUCUS. The regular movements of the stomach churn up the food and gastric juice to a thick fluid called chyme.

When digestion in the stomach is complete the pyloric sphincter relaxes and allows a small amount of the chyme into the duodenum.

Small intestine

Duodenum
The first part of the small intestine is called the duodenum. Chyme passing into the duodenum causes the pancreas to produce pancreatic juice which contains sodium bicarbonate. This juice makes the chyme alkaline again. Pancreatic juice also contains several powerful enzymes. It is in the duodenum that most of digestion takes place. The pancreas is also an endocrine gland (see chapter 35).

Chyme in the duodenum also causes BILE to pass down the bile DUCT from the gall bladder in the liver. The bile is a green substance which contains certain salts that cause fats to emulsify, which means to break up into tiny droplets. This allows the fat-digesting enzymes to work. Bile also contains sodium bicarbonate which helps produce the alkaline medium needed for digestion in the duodenum.

Ileum
The next part of the small intestine is called the ileum. The intestinal juices (succus entericus) contain many enzymes which complete the process of digestion to produce the soluble products which can be ABSORBED into the blood stream.

The whole of the inside surface of the ileum is covered by finger-like projections called VILLI which are about 1 mm long. This gives the inside of the ileum a much greater SURFACE AREA than if it had a smooth lining (see diagram 3). Inside each villus is a network of blood VESSELS and a single LACTEAL or LYMPH vessel. The digested food passes into these vessels and is transported to the liver and around the body.

Large intestine

At the point where the ileum joins the colon is the appendix. In HERBIVORES, such as rabbits, this is large and contains bacteria which help digest CELLULOSE. In man, the appendix is small and does not seem to have a function.

The large intestine consists of the colon, rectum and anus. The colon receives indigestible material from the ileum. This remains in the colon for about thirty-six hours during which time most of the water and salts are re-absorbed. The semi-solid waste material called FAECES is removed from the body via the rectum and through the anus.

In man the alimentary canal is about ten metres long from mouth to anus.

Related reading

Chapter 12, Enzymes
Chapter 13, The liver

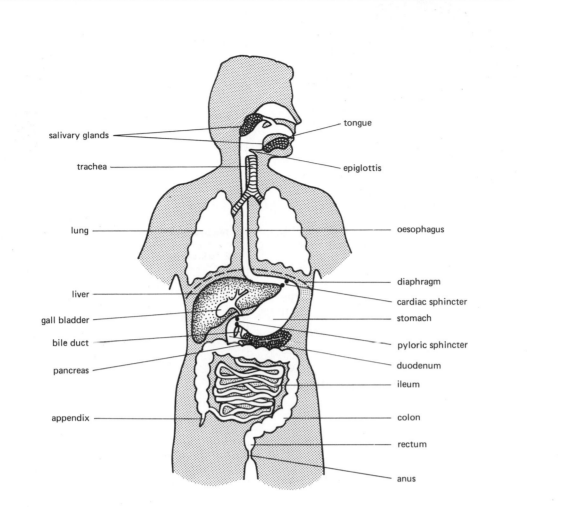

salivary glands

tongue

trachea

epiglottis

lung

oesophagus

liver

diaphragm

gall bladder

cardiac sphincter

bile duct

stomach

pancreas

pyloric sphincter

appendix

duodenum

ileum

colon

rectum

anus

Diagram 1   Alimentary canal of man

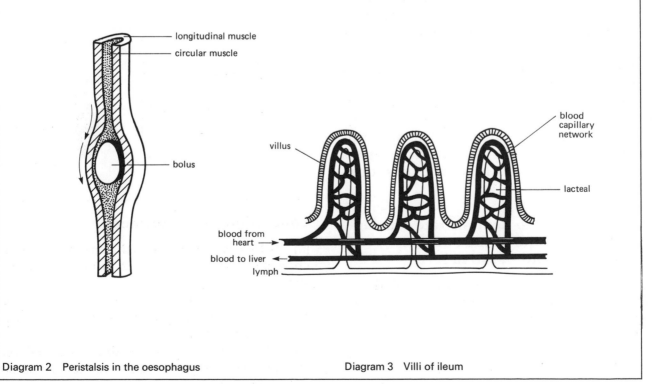

longitudinal muscle

circular muscle

blood capillary network

villus

bolus

lacteal

blood from heart

blood to liver

lymph

Diagram 2   Peristalsis in the oesophagus

Diagram 3   Villi of ileum

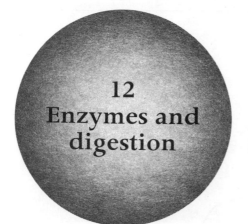

12
Enzymes and digestion

Food must pass into the cells of any living ORGANISMS, because it is in the cells that food is needed to give energy to do work, and to give materials to build new TISSUE.

Carbohydrates, fats and proteins are large MOLECULES and cannot pass across the cell MEMBRANE. Digestion breaks carbohydrates into glucose, fats into fatty acids and proteins into amino acids. Glucose, fatty acids and amino acids are smaller molecules which can cross the cell membrane and enter the cell.

The need for enzymes

Starch can be broken down into smaller molecules in a test tube using high temperatures and strong acid. This process cannot be used inside the body without damaging or killing the organism. However, starch chewed with saliva in the mouth is quickly broken down into the smaller molecules of maltose. This happens because of the presence of the enzyme amylase in the saliva. ENZYMES are substances which speed up chemical changes without the need for high temperatures or strong acids.

Enzymes are found throughout the digestive systems of animals and are also present inside all living cells. These chemicals are important inside cells because they speed up the release of energy from glucose, and speed up the building of new tissue. There are many different enzymes in the living world but they all have some CHARACTERISTICS in common.

An experiment to demonstrate some of the characteristics of the enzyme AMYLASE found in saliva

Information needed: When iodine solution is added to a solution containing starch there is a colour change from brown to blue-black.

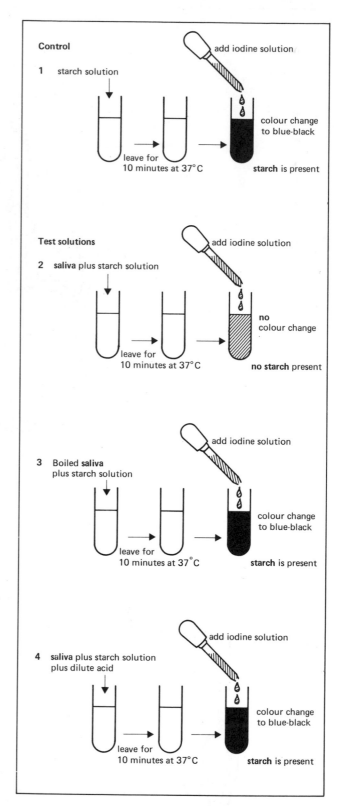

Conclusions to experiment

(a) Saliva, which contains the enzyme amylase, changed starch in tube 2 into a different substance.

(b) Heated enzyme in tube 3 had no action because enzymes are proteins and are denatured by high temperatures.

(c) The enzyme in tube 4 did not speed the reaction because the acidity was too great for amylase to act.

The characteristics of enzymes

1 All enzymes are proteins and because of this they can be broken down by EXCESS heat. This is called DENATURING the enzyme.

2 Enzymes are made by living cells to speed up chemical changes. They are sometimes called biological catalysts.

3 Enzymes are specific. This means that for each chemical reaction there is a particular enzyme to speed up that particular reaction.

4 Small amounts of enzymes are needed for any specific chemical change.

5 Enzymes work most efficiently at an ideal pH. pH is a measure of acidity and alkalinity. If conditions are too strongly acid or too strongly alkaline for any particular enzyme, it will not speed up the reaction.

6 Enzymes work most efficiently at an ideal temperature. The digestive enzymes of man work best at body temperature, which is 37°C.

There are many experiments which demonstrate the characteristics of enzymes. Experiments with amylase are shown opposite.

Enzymes in the human alimentary canal

Each part of the alimentary canal of man has particular enzymes which break down large molecules. The final products of digestion can be taken into the blood and LYMPHATIC system from the small intestine and carried to each cell in the body. Water is essential for digestion. MUCUS is added to help the food move along the gut and to prevent the enzymes digesting the walls of the gut itself. Details of the enzymes in the human alimentary canal are shown in the table.

Enzymes in the human alimentary canal

Part of the alimentary canal	What happens to CARBOHYDRATES	What happens to FATS	What happens to PROTEINS	Extra information
MOUTH	Enzyme – AMYLASE Change – starch to maltose			Digestive juice – SALIVA. pH – alkaline. Water and mucus added.
STOMACH			Enzyme – PEPSIN Change – proteins to peptones Enzyme – RENNIN Change – milk protein to curds	Digestive juice – gastric juice. pH – acid. Mucus and hydrochloric acid added.
DUODENUM	Enzyme – AMYLASE Change – starch to maltose	Enzyme – LIPASE Change – fats to fatty acids	Enzyme – TRYPSIN Change – peptones to peptides	Digestive juice – pancreatic juice. pH – alkaline. Bile added from gall bladder.
ILEUM (Small intestine)	Enzyme – MALTASE Change – maltose to glucose	Enzyme – LIPASE Change – fats to fatty acids	Enzyme – PEPTIDASE Change – peptides to amino acids	Digestive juice – succus entericus. pH – alkaline GLUCOSE, FATTY ACIDS, and AMINO ACIDS ABSORBED into blood.
COLON				4/5th of water in the solid waste is REABSORBED

Related reading

13
Food storage

Animals and plants must store food in their bodies so that they can still live during food shortages. Food is also needed for a developing EMBRYO such as a seed or an egg.

Food storage in Man

When digestion is complete the food is in the form of glucose, amino acids or fatty acids. These are soluble and so must be converted to an insoluble form if they are to be stored (see diagram 1).

Glucose

The glucose is carried by the hepatic portal vein from the ileum to the liver, and then carried to the cells of the body in the blood stream. Any glucose in excess of 80 mg per 100 cm³ of blood is converted to insoluble glycogen which is then stored in the liver or in the muscles.

The stored glycogen in the liver can be converted back to glucose when the blood-sugar level drops below 80 mg/100 cm³.

Some glucose can be converted to fat for storage. The liver also stores certain vitamins and MINERALS.

Amino acids

The amino acids are also carried by the hepatic portal vein to the liver. Those that are needed to make new TISSUE are carried to the cells by the bloodstream. However, EXCESS amino acids cannot be stored. They must first have their nitrogen removed (DEAMINATION) and then they can be stored as glycogen or fat. The removed nitrogen is in the form of ammonia which is changed to urea in the liver and then transported by the blood to the excretory organs.

Fatty acids

Most of the fatty acids are re-formed into fats and enter the LACTEALS. The fat is either used for energy or stored in fat cells beneath the skin or around organs such as the kidneys.

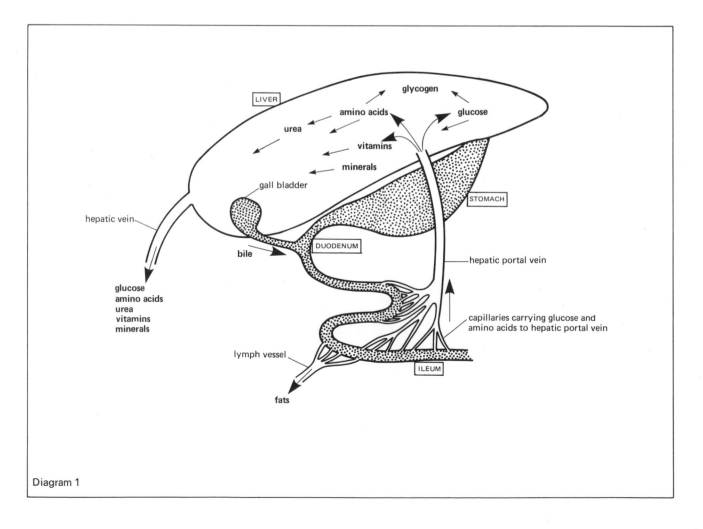

Diagram 1

Food storage in plants

Food is necessary to the plant for growth and respiration. When excess food is made it can be transported in the PHLOEM to some part of the plant for storage. Sugars, which are made during PHOTOSYNTHESIS, are converted to starch and stored in the cells for use during the night.

The form in which food is stored can be carbohydrate, fat or protein depending on the SPECIES of plant and the part of the plant which is used for storage.

Types of food storage

(i) **Stem.** In sugar cane the whole stem is swollen due to the storage of SUCROSE.

(ii) **Corm.** (See diagram 2) In crocuses the base of the stem is swollen with stored food. This corm stays below ground.

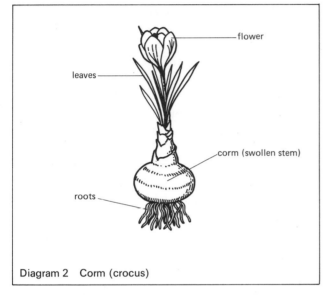

Diagram 2   Corm (crocus)

(iii) **Bulb.** (See diagram 3) In onions or lilies the base of the leaves are swollen with stored food. This bulb stays below ground.

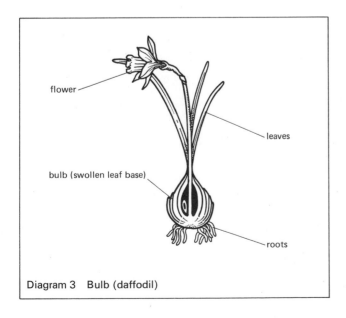

Diagram 3   Bulb (daffodil)

(iv) **Rhizome.** (See diagram 4) In iris or couch grass the horizontal underground stem becomes swollen with stored food. This rhizome stays below ground and lasts for several years.

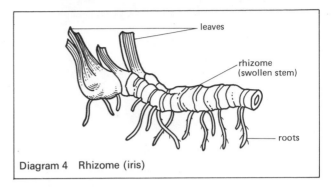

Diagram 4   Rhizome (iris)

(v) **Stem tuber.** (See diagram 5) In potatoes or yams, only the tips of the underground stem become swollen with stored food to form the tuber.

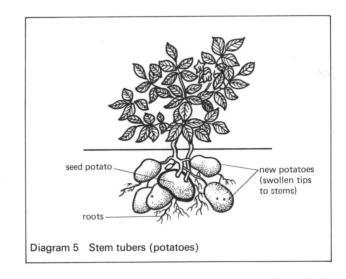

Diagram 5   Stem tubers (potatoes)

(vi) **Tap root.** In carrots or sugar beet, the whole root becomes swollen with stored food.

(vii) **Tuberous root.** In dahlias or sweet potatoes, the roots become swollen with stored food and resemble the stem tubers.

(viii) **Seeds.** In the seeds of peas and beans the food is stored in special seed leaves called COTYLEDONS. In cereals such as maize the food is stored in the endosperm of the seed.

Many plants use food storage as a way of surviving the winter months. In the spring, rapid growth and flower formation take place. Often buds develop on the food store. This is a method of asexual reproduction or vegetative reproduction.

Related reading

Chapters 11 and 12, Products of digestion
Chapter 48, Asexual reproduction in plants
Chapter 51, Structure of seeds

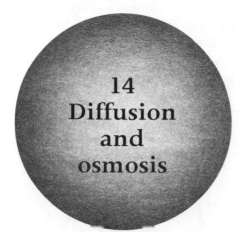

14
Diffusion and osmosis

Water can exist in three forms: a gas which is steam, a liquid which is water, and a solid which is ice. MOLECULES or particles are believed to be moving at all times in all substances. Molecules move very fast in gases, less fast in liquids and most slowly in solids. Therefore water molecules moving very fast make up the gas called steam. Water molecules moving less fast make up the liquid called water. The slowest moving molecules make up the solid called ice. This continual movement means that different kinds of molecules gradually mix together without any stirring or shaking. This is called DIFFUSION.

Diffusion

An example of diffusion is found when a cube of sugar is carefully placed in a beaker of water and left for several days. The sugar gradually dissolves in the water and becomes evenly spread throughout the liquid. The brightly coloured crystals of potassium permanganate can be used to show this in the same way as sugar. In these crystals the colour shows the gradual diffusion of the coloured substance over a period of days (see experiment 1).

In the two examples used, sugar and potassium permanganate are called the SOLUTES as they are the solids and water is called the SOLVENT. There are many kinds of other solvents such as the grease solvents used in dry cleaning. However, in biological studies, water is the most common solvent.

Diffusion takes place faster in gases because the molecules or particles move more quickly than in liquids or solids. The scent of roses is detected by the nose because particles from the rose gradually move by diffusion away from the flowers. Plants take carbon dioxide into the leaves by diffusion. Oxygen is taken into the human blood from the air in the lungs by diffusion.

Liquids diffuse into their surroundings by evaporating, such as the water lost in TRANSPIRATION or from the sea to form clouds.

Solid particles of smoke and dirt diffuse into the atmosphere and cause pollution miles away from industrial areas.

In diffusion, the movement of particles is always away from a place of higher concentration or strength to a place of lower concentration or weakness.

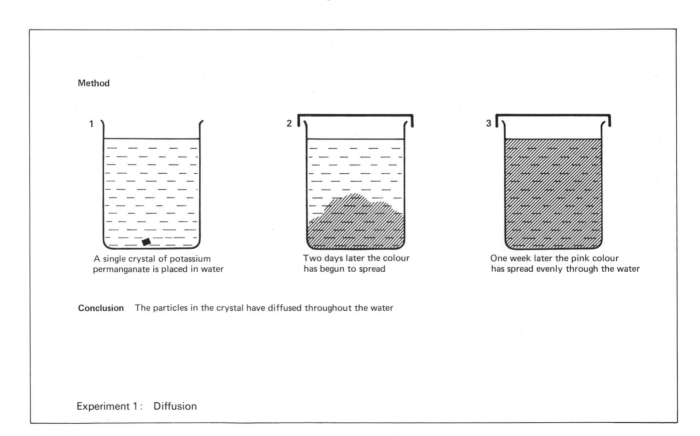

Method

1 A single crystal of potassium permanganate is placed in water

2 Two days later the colour has begun to spread

3 One week later the pink colour has spread evenly through the water

Conclusion   The particles in the crystal have diffused throughout the water

Experiment 1:   Diffusion

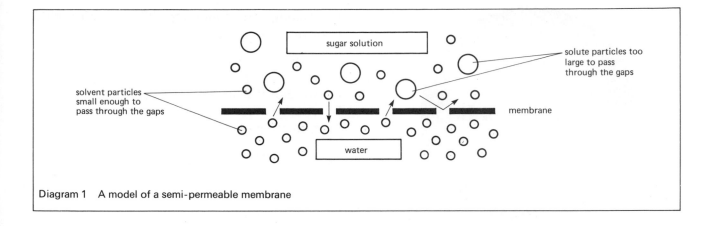

Diagram 1   A model of a semi-permeable membrane

Osmosis

The controlled flow of water into and out of the cell is called OSMOSIS. This flow is necessary because most living cells are about 80% water. Land animals and plants are constantly losing water to the air by diffusion. Plants transpire and animals must excrete wastes. Freshwater animals and plants are in constant danger of taking in too much water by diffusion from their surroundings and so bursting their cells.

Osmosis is a special case of diffusion by the solvent water. Osmosis is possible in living cells because they are surrounded by a SEMI-PERMEABLE MEMBRANE (see diagram 1). This membrane allows solvent molecules to pass across the membrane, but does not allow the larger solute molecules to pass out of the cell. This means that water can diffuse from the soil

into the cells of the root hairs and so on into each cell of the plant right up to the leaves. Too much water is prevented from entering the cell by the build-up of pressure against the cell wall. This is called osmotic pressure.

Osmosis can be demonstrated in living TISSUE using potato tissue (see experiment 2). Boiling one half of the experimental potato shows that destroying the cell membrane destroys the ability of the tissue to take up water by osmosis.

Related reading

Chapter 4, Cell structure
Chapter 21, Transport in plants
Chapter 24, Water regulation in animals
Chapter 25, Water regulation in plants
Chapter 68, Control experiments

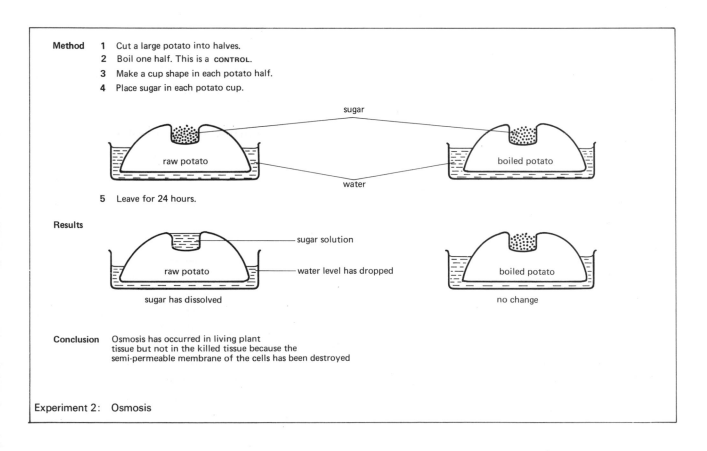

Experiment 2:   Osmosis

Questions B

The answers to questions 1 to 7 are shown by one of the letters A, B, C, D or E.

1 Of the following organisms found in a habitat, the one which makes food from simple substances is the:

A toadstool.
B moss.
C ant.
D earthworm.
E greenfly.

2 Of the following, the food which yields most energy per kilogram is:

A butter.
B bread.
C biscuits.
D peas.
E beef.

3 When a dog is feeding, the main function of its incisor teeth is to:

A cut.
B tear.
C crush.
D masticate.
E grind.

4 Bile emulsifies fats, which means that the fat is made into:

A small lumps.
B a solid.
C droplets.
D a solution.
E molecules.

5 The function of the enzyme salivary amylase is to:

A digest fatty acids.
B convert starch into maltose.
C make stomach liquids alkaline.
D emulsify fats.
E cause milk to clot.

6 Among the following, the activity NOT performed by the liver is:

A formation of bile.
B storage of glycogen.
C storage of iron.
D storage of amino acids.
E formation of urea.

7 The passage of water through a semi-permeable membrane is called:

A osmosis.
B transpiration.
C fermentation.
D anaerobiosis.
E synthesis.

8 Copy out and then complete the following sentences:
(i) Pepsin and renin are both called ....
(ii) Final products of the digestion of protein in the small intestine are called ....
(iii) The food storage organ of a crocus is called
....
(iv) Food is moved along the alimentary canal by a muscle action called ....
(v) The number of teeth in the adult human mouth should be ....
(vi) Animals which eat both meat and plant materials are called ....

9 Here is a typical pyramid of numbers:

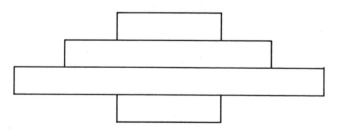

Here is the outline of an unusual pyramid of numbers:

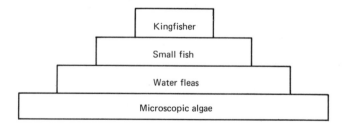

Draw the outline above making a likely pyramid of numbers using FOUR of the organisms listed below:

rose trees   ladybirds
whales   human beings
sparrows   aphids (greenflies)
antelopes   plankton

10 Write about six lines on FIVE of the following:

(i) Animals as consumers
(ii) A semi-permeable membrane
(iii) Bulbs and corms
(iv) Food webs
(v) Amino acids
(vi) Saprophytes
(vii) Roughage
(viii) Swallowing
(ix) The pancreas
(x) Pyramids of numbers

11 Write an account of about twenty-five lines on ONE of the following. (Diagrams may improve your answer.)

(i) Digestive enzymes in man
(ii) The liver as a storage organ
(iii) The small intestine
(iv) Diffusion
(v) Structure of a tooth

12

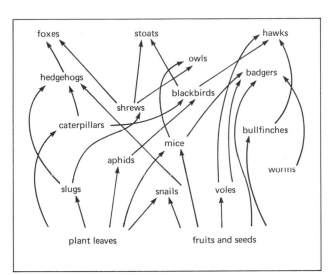

Use this food web to answer the following:
(i) Name two organisms eaten by hawks.
(ii) Write a food chain from this food web involving more than three organisms.
(iii) Name one omnivore in the web.
(iv) Suggest a habitat where this web may have developed.

(SREB)

13 In an experiment on the effects of temperature on enzyme activity, a solution of amylase was used to digest starch at different temperatures.
The following results were obtained:

Temperature (°C)	10	20	30	40	45	50
Time for starch digestion (minutes)	20	9	5	2	7	25

(i) Draw a graph to show these results on graph paper.
(ii) Describe the effect of temperature on enzyme activity as shown by this experiment.
(iii) Using your graph state:
(a) the time it will take for the starch to be digested at 25°C.
(b) the temperatures at which it will take 4 minutes for the starch to be digested.
(iv) What apparatus would you use to maintain the different temperatures?
(v) How would you test for starch and what result would you expect if starch were present?

(ALSEB)

14 In the following table, all weights relate to the amount of food contained in 25 g.
(Kilojoules are units used instead of Calories.)

Food	Kilojoules per 25 g	Protein g	Fat g
White bread	289	2.2	0.4
Butter	945	0.1	24.2
Cheese	500	7.2	9.8
Potatoes	105	0.6	0
Blackcurrant	33	0.3	0

Food	Carbo-hydrate g	Calcium mg	Vitamin C mg	Vitamin D mg
White bread	14.9	26	0	0
Butter	0	4	0	11
Cheese	0	230	0	4
Potatoes	5.9	2	2.5	0
Blackcurrant	1.9	17	57	0

(i) From the above table answer the following questions:
(a) Which food is richest in protein?
(b) Which food is of least value in bone formation?
(c) Which food would you recommend to someone suffering from scurvy?
(ii) What do joules (calories) measure?
(iii) How many kilojoules would you obtain if you ate a cheese sandwich containing 100 g of bread, 50 g of cheese and 2.5 g butter?

(Met. REB)

15 (i) What is osmosis?
(ii) Make a labelled diagram of the apparatus you would use to show the process of osmosis. What results would you expect and how would you explain them?

(MREB)

15
Respiration

Internal Respiration

Food contains stored energy in the form of complex compounds. These compounds are manufactured by green plants from simple substances such as carbon dioxide and water using the sun's energy by PHOTOSYNTHESIS. Animals and plants recover the energy stored in the complex compounds by breaking them down into simple substances. This is a process which takes place inside the living cell and is called respiration, or sometimes internal respiration. Respiration can be shown as an equation:

$$\text{glucose} + \text{oxygen} \longrightarrow \text{carbon dioxide} + \text{water} + \text{energy}$$

Photosynthesis can be shown as a similar equation which moves in the opposite direction:

$$\text{carbon dioxide} + \text{water} + \overset{\text{light}}{\underset{\text{energy}}{}} \longrightarrow \text{glucose} + \text{oxygen}$$

Thus respiration is the recovery of the stored energy in food by breaking down that food, inside the living cells, into simple chemicals.

Aerobic respiration

Aerobic respiration is a breakdown of food using oxygen, which is similar to the breakdown of petrol or wood by burning. These fuels break down, in combustion with oxygen, to give carbon dioxide, water vapour and heat energy. Living ORGANISMS cannot withstand high temperatures so they break down their food step by step to produce chemical energy. This chemical energy is available to do the work of the whole organism.

Respiration using oxygen can be thought of as taking place in two stages. Both stages take place inside living cells:

Stage One Glucose is broken down into simpler compounds using ENZYMES. This stage takes place inside the CYTOPLASM of the cell.
Stage Two The products of stage one are transported to special ORGANELLES inside the cell called MITOCHONDRIA. Here, enzymes complete the breakdown to carbon dioxide and water.

The amount of energy released by complete breakdown of 10 GRAMS of glucose to carbon dioxide and water is about 157 KILOJOULES.

Most living organisms do not survive in the absence of oxygen as the products of stage one are poisonous and quickly kill the organism. However, some simple organisms, such as bacteria and fungi, can survive for long periods without oxygen. Also muscle cells of many animals, including man, can respire without oxygen for a limited time if oxygen is in short supply.

Anaerobic respiration

Anaerobic respiration is the releasing of energy from food in the absence of oxygen. This is achieved in some living cells by using only stage one of respiration. This releases much less energy than aerobic respiration. For example 10 grams of glucose produces only about 6.5 kilojoules of energy.
Plants In plants, like yeast, the poisonous products of stage one are changed into carbon dioxide and alcohol. These waste products are well known to man as the alcohol of beer and wine making, and as the bubbles of carbon dioxide which cause bread to rise.

$$\text{glucose} \longrightarrow \text{carbon dioxide} + \text{alcohol} + \text{energy}$$

This can be called fermentation.

Certain plant seeds can respire anaerobically for short periods. But the alcohol and carbon dioxide produced eventually kills the seed, if these are not constantly removed from the surroundings. Plant seeds will not germinate in the absence of oxygen because the energy requirements for such rapid growth are greater than the energy available from anaerobic respiration.
Animals In animals, like man, muscle cells suffer a shortage of oxygen at times. This can be brought about either by long periods of exercise or short periods of intense activity. The muscle cells overcome this lack of oxygen by using only stage one of respiration. The poisonous products of stage one are changed into lactic acid in the cells. However, too much lactic acid produces the pain known as cramp.

The experiments shown opposite demonstrate aerobic and anaerobic respiration.

Related reading

Chapter 4, Cells
Chapter 12, Enzymes
Chapter 68, Control experiments

Demonstrations of respiration
Information needed: (i) Lime water changes from a clear liquid to a cloudy liquid when carbon dioxide gas is bubbled through
(ii) Carbon dioxide is removed from the air by sodium hydroxide solution

1 AEROBIC RESPIRATION

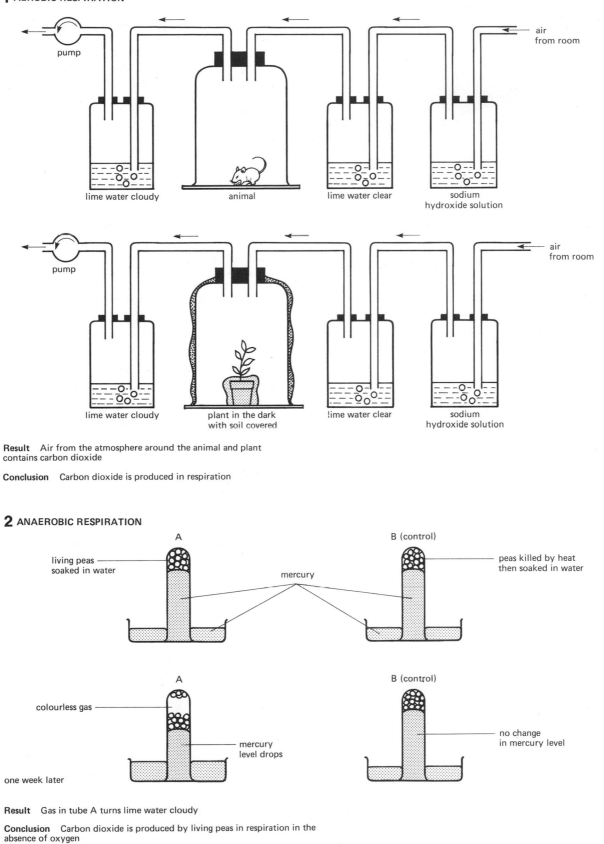

lime water cloudy   animal   lime water clear   sodium hydroxide solution

pump   air from room

lime water cloudy   plant in the dark with soil covered   lime water clear   sodium hydroxide solution

pump   air from room

Result Air from the atmosphere around the animal and plant contains carbon dioxide

Conclusion Carbon dioxide is produced in respiration

2 ANAEROBIC RESPIRATION

A   B (control)

living peas soaked in water   mercury   peas killed by heat then soaked in water

A   B (control)

colourless gas   mercury level drops   no change in mercury level

one week later

Result Gas in tube A turns lime water cloudy

Conclusion Carbon dioxide is produced by living peas in respiration in the absence of oxygen

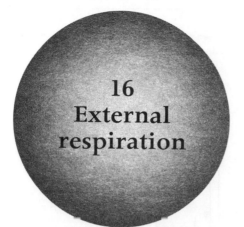

16
External respiration

Most living ORGANISMS use oxygen inside the body cells in the process of energy release from food. This process is called internal respiration. The energy release produces the waste products carbon dioxide and water. These waste products have to be removed from the cell back into the surroundings. The process of taking oxygen from the surroundings into the cells and getting rid of waste carbon dioxide and water vapour is called external respiration.

Simple water animals and plants usually manage external respiration by DIFFUSION directly from the water to the cells. Carbon dioxide dissolves readily in water, and oxygen is also slightly soluble.

Large water animals, such as fishes, have special parts of the body developed as gills where diffusion takes place into the blood system (see diagram 1). The blood then carries the oxygen to the cells, and the carbon dioxide back to the gills and so into the water.

Land animals must take oxygen from the air which contains approximately twenty per cent oxygen and seventy-nine per cent nitrogen. The air returned from the lungs of a human being has less oxygen and has more carbon dioxide and water vapour than the atmosphere.

Approximate composition of inhaled and exhaled air.

Gases	INHALED AIR (Atmospheric air)	EXHALED AIR (Air from the lungs)
OXYGEN	20%	16%
NITROGEN	79%	79%
CARBON DIOXIDE	0.04%	4%
WATER VAPOUR	a trace	SATURATED
RARE GASES	1.0%	1.0%

Living cells are about 80% water and if they were exposed to the dry air they would quickly lose water by diffusion to the atmosphere. For this reason, land animals and plants have a protective covering of waterproof material. Air must be taken inside the body and brought into contact with the cells without the cells being in direct contact with the dry air.

Land-dwelling arthropods, such as insects, have a system of tubes through which air and waste carbon dioxide pass. These tubes are called

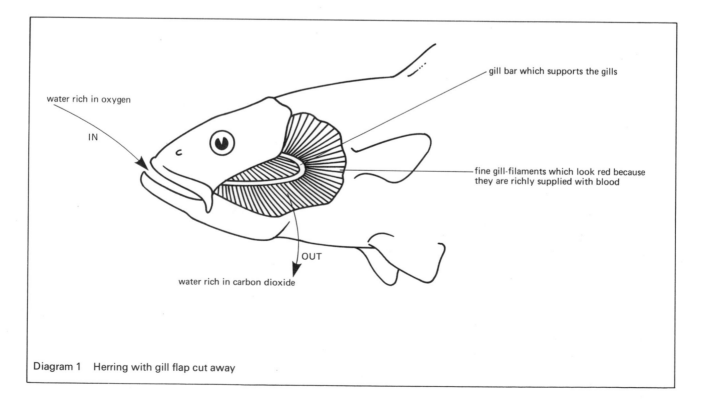

water rich in oxygen

IN

gill bar which supports the gills

fine gill-filaments which look red because they are richly supplied with blood

OUT

water rich in carbon dioxide

Diagram 1   Herring with gill flap cut away

tracheae and they have openings on the side of the insect called spiracles. Blood plays no part in carrying the gases around the insect body as the tracheae branch to all parts of the animal. The gases pass into the cells by diffusion (see diagram 2).

Plants respire through openings in the leaves called STOMATA. These openings allow diffusion to take place in the damp conditions which are present because of TRANSPIRATION.

External respiration in man is discussed fully in chapter 17.

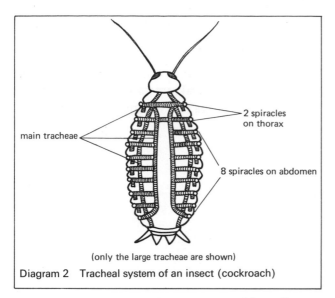

main tracheae

2 spiracles on thorax

8 spiracles on abdomen

(only the large tracheae are shown)

Diagram 2   Tracheal system of an insect (cockroach)

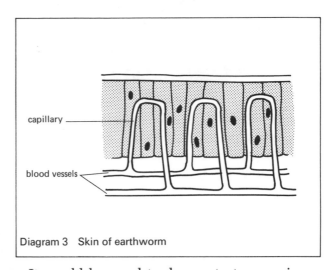

capillary

blood vessels

Diagram 3   Skin of earthworm

Animals such as earthworms are able to live on land only by keeping the surface of the body moist at all times. This is done by the animal secreting MUCUS over the whole body and spending most of its life in the damp earth. Oxygen and carbon dioxide can therefore diffuse across the whole surface of the body. The gases are carried to the body cells by blood vessels found close to the surface of the skin (see diagram 3).

It would be cruel to demonstrate experimentally that animals die in the absence of oxygen. However, the experiment below shows that oxygen is needed for germination. As seeds must respire aerobically before they can germinate, the experiment also indicates that oxygen is needed for aerobic respiration.

Related reading

Chapter 15, Internal respiration
Chapter 17, Breathing in man
Chapter 68, Control experiments

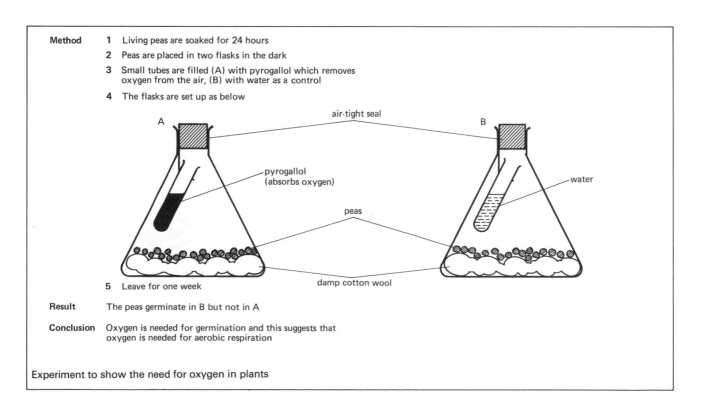

Method	1	Living peas are soaked for 24 hours
	2	Peas are placed in two flasks in the dark
	3	Small tubes are filled (A) with pyrogallol which removes oxygen from the air, (B) with water as a control
	4	The flasks are set up as below

air-tight seal

A

B

pyrogallol (absorbs oxygen)

water

peas

damp cotton wool

5   Leave for one week

Result   The peas germinate in B but not in A

Conclusion   Oxygen is needed for germination and this suggests that oxygen is needed for aerobic respiration

Experiment to show the need for oxygen in plants

17
Breathing in man

In man, and the other mammals, the exchange of gases takes place at a special region called a respiratory surface, which is part of a respiratory ORGAN called the lungs.

Thorax

The lungs and heart of man are situated in the THORAX (see diagram 1). The walls of the thorax are strengthened by the ribs which also act as protection. The floor of the thorax is a muscle sheet called the diaphragm. The rib cage is airtight, and so movements of the diaphragm change the volume of the thorax. This change in volume results in a change of pressure which enables the lungs to expand or be compressed. The lungs are surrounded by a slippery, double skin called the pleural MEMBRANES which reduce friction as the lungs rub against the thorax wall.

Diagram 1   The thorax

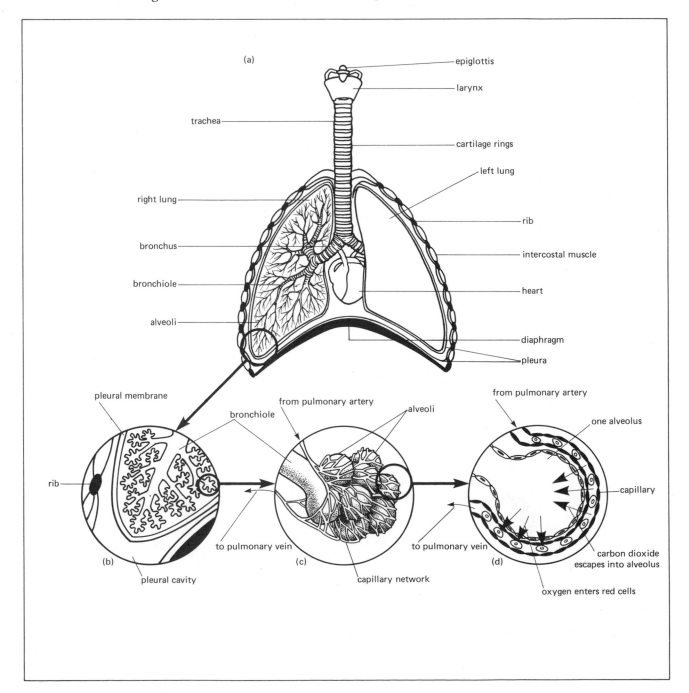

Inhalation (breathing in)

When the diaphragm contracts it is pulled down and flattened. At the same time the intercostal muscles contract, pulling the ribs upwards and outwards (see diagram 2). These actions cause an increase in the volume of the thorax, and so reduce the internal pressure. This causes air to rush in through the nose and mouth.

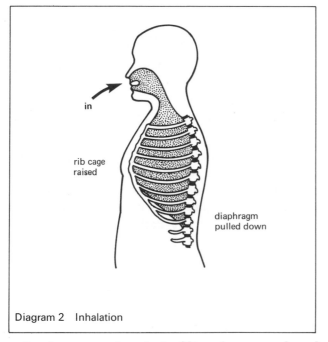

Diagram 2   Inhalation

In the nose, the air is filtered, warmed and moistened. It then passes over the vocal cords in the larynx and into the trachea, which is strengthened with rings of CARTILAGE. The air flows into the left or right bronchus which then further divides into bronchioles. These bronchioles end in air sacs called alveoli (see diagram 1b).

Gaseous exchange

There are about 700 million alveoli in a human's lungs. The structure of the alveoli gives a large internal SURFACE AREA which is covered by a fine network of blood vessels called CAPILLARIES (see diagram 1c). These capillaries are so fine that the red cells in the blood must slowly squeeze through them. This allows the oxygen to DIFFUSE from the air in the lungs through the alveoli walls into the red cells in the blood (see diagram 1d). The oxygenated blood then passes into the pulmonary vein and on to the heart for circulation around the body.

The pulmonary artery brings deoxygenated blood from the heart into the capillaries. This blood contains a high concentration of carbon dioxide which diffuses from the blood, through the alveoli wall, into the air in the lungs. This air is removed during exhalation.

Exhalation (breathing out)

This is the reverse of inhalation. The intercostal muscles relax allowing the rib cage to drop under its own weight (see diagram 3). The diaphragm relaxes and returns to its original dome-shape. This reduces the volume in the thorax and so increases the internal pressure. The air is forced out of the lungs.

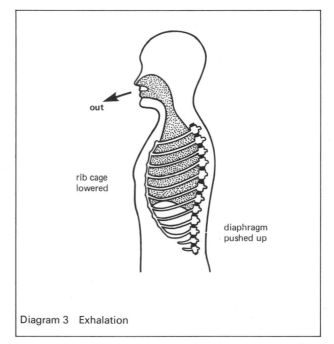

Diagram 3   Exhalation

The exhaled air will contain more carbon dioxide and water, but less oxygen than the inhaled air.

Lung capacity

The average capacity of the lungs for an adult human is about five litres but in normal breathing only about half a litre of air passes in and out of the lungs. This is called the tidal air.

The thorax cannot collapse completely and so about one litre of air can never be expelled from the lungs. This is called residual air. Residual air does not go stale or stagnant as it is mixed with tidal air during every breath.

The total amount of air that can be forced out of the lungs after a deep breath is known as a person's vital capacity. The vital capacity is usually about four litres.

Related reading

Chapter 16, External respiration
Chapter 18, Transport of oxygen in man
Chapter 22, Healthy lungs

18
Transport in mammals

Food from the gut, and oxygen from the lungs must reach every cell in the body of a mammal. Many of these cells are deep inside the animal and must have the essential substances carried to them in some way. This need is met by a system of tubes containing a fluid called blood. The system is kept moving by a pump called the heart.

The structure and function of blood

Blood placed in a tube which contains a citrate solution will not clot. When the tube has been left to settle the blood separates into layers (see diagram 1). The red part at the bottom is made up of red cells. A thin white layer on top of the red contains the platelets and white cells. The straw coloured fluid is plasma. The blood is eighty per cent water and twenty per cent solids.

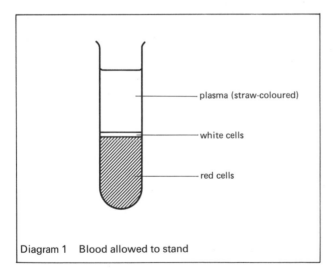

Diagram 1   Blood allowed to stand

Red cells

These cells have no NUCLEI and are of a fixed shape (see diagram 2a). Inside the cell is haemoglobin which is a chemical containing iron which combines with oxygen. Each cubic millimetre of human blood contains approximately five million red cells. The average human adult man has about six litres of blood.

This makes a total of thirty million, million, red cells travelling all around the body carrying oxygen to every single cell. They also carry some carbon dioxide back to the lungs in the opposite direction. Red cells are continually made in the bone marrow, and last for only three months in the blood system.

White cells

These cells are larger than red cells. They have a nucleus and have no fixed shape (see diagram 2b). There are two different types:
(i) the lymphocytes which make chemicals which kill bacteria
(ii) the polymorphs which are bigger cells which flow around a bacterium, or foreign body, and destroy it.
Each cubic millimetre of blood contains about 8000 white cells.

Both of these types of cell can move about by flowing along and can squeeze through the walls of blood VESSELS to attack the invading germs. White cells are sometimes called the soldiers of the body because they gather around a wound and attack the germs. Dead white cells and dead bacteria are seen in a wound as pus.

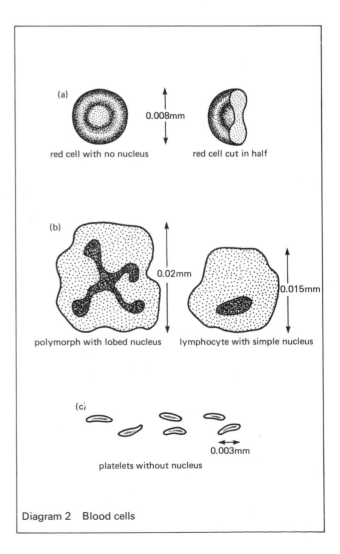

Diagram 2   Blood cells

Platelets

These are small fragments of cell material. They have no nucleus and are irregular in shape (see diagram 2c). The job of the platelets is to help in clotting when a blood vessel is broken, so stopping blood from leaking out.

Plasma

This is a yellow-coloured liquid which is made up of water with dissolved chemicals. The chemicals are carried from place to place all over the body. They include dissolved food from the gut to the cells; waste materials from the cells to the kidneys and skin; carbon dioxide from the cells to the lungs, and HORMONES from all the endocrine glands to the places where they act.

MINERAL SALTS of calcium, sodium and of many other elements are also in the plasma.

Plasma proteins are needed to make a blood clot with the platelets.

Body temperature

Mammals and birds are warm-blooded animals. This warmth is the result of muscle activity and chemical activity inside cells. The blood spreads this heat and helps to keep all parts of the body at the same temperature.

Making a blood smear (see diagram 3)

Blood cells can be studied by spreading blood on a microscope slide and staining it with dyes. Each type of cell can be seen at a reasonable magnification. Great care should be taken to ensure that each lancet is used only once, due to the serious risk of transmitting infectious blood diseases.

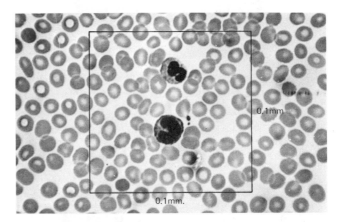

0.1mm.

0.1mm.

Microphotograph of blood smear

Related reading

Chapter 19, Circulation of blood, blood groups and clotting

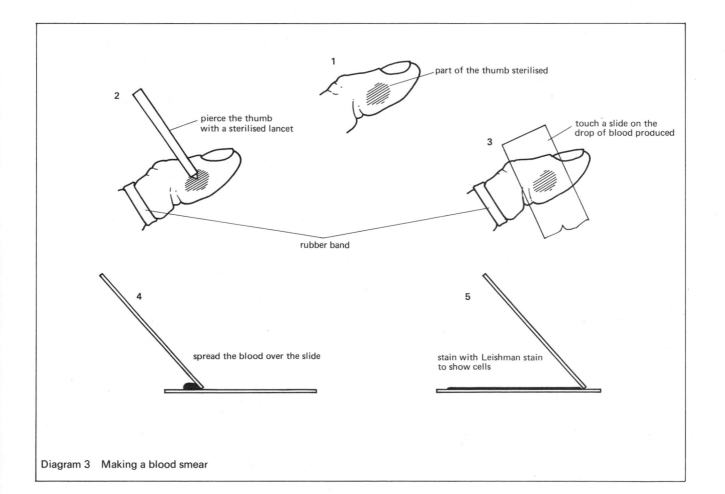

1 part of the thumb sterilised

2 pierce the thumb with a sterilised lancet

3 touch a slide on the drop of blood produced

rubber band

4 spread the blood over the slide

5 stain with Leishman stain to show cells

Diagram 3 Making a blood smear

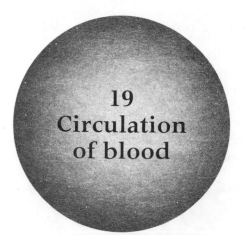

19
Circulation of blood

Blood is a fluid which is carried around the body in a system of tubes called blood VESSELS. There are three main types of blood vessels called arteries, CAPILLARIES and veins (see diagram 1).

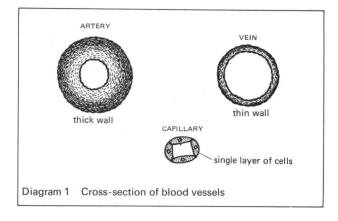

Diagram 1   Cross-section of blood vessels

Arteries

Arteries are thick-walled, muscular, elastic vessels carrying oxygenated blood away from the heart at high pressure. This is true of all arteries except the pulmonary artery which carries deoxygenated blood to the lungs.

A pulse can be felt in an artery near the surface of the skin, such as in the wrist or neck. This is caused by the contraction of the heart and the elasticity of the artery walls.

Arteries divide into smaller blood vessels called arterioles. Arterioles divide many times to form a network of capillaries.

Capillaries

Capillaries are so fine that they have walls only one cell thick. They form a dense network all over the body, so that every cell of the body receives its requirements from a capillary that will never be more than a fraction of a millimetre away. All the functions of the blood depend on this network of capillaries.

Capillaries join together to form wider vessels called venules. The venules join together to form veins.

Veins

Veins are thinner walled, less muscular and less elastic than arteries. They carry deoxygenated blood to the heart at low pressure. Because of this low pressure, most veins have valves (see diagram 2) which prevent the back-flow of blood.

The pulmonary vein is an exception to this definition as it carries oxygenated blood from the lungs to the heart.

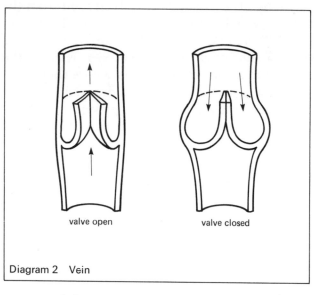

Diagram 2   Vein

Unusual features of certain blood vessels (see diagram 3)

Pulmonary artery
This carries deoxygenated blood and has a high carbon dioxide content.

Pulmonary vein
This carries oxygenated blood and has a low carbon dioxide content.

Hepatic portal vein
This has a very high food content as it goes from the intestine to the liver. It does not transport blood from an organ to the heart like the other veins. It both begins and ends in capillaries.

Renal vein
This has a very low urea (waste product) content as the blood has come from the kidneys which remove the waste products.

Lymphatic system

LYMPH is a fluid which DIFFUSES out of the blood into the spaces outside the capillaries. It is similar to blood plasma, but contains less protein. The lymph diffuses into lymph capillaries which join up to form larger lymphatics. Two larger lymphatic ducts return the collected lymph into the bloodstream at the base of the neck, in the jugular vein.

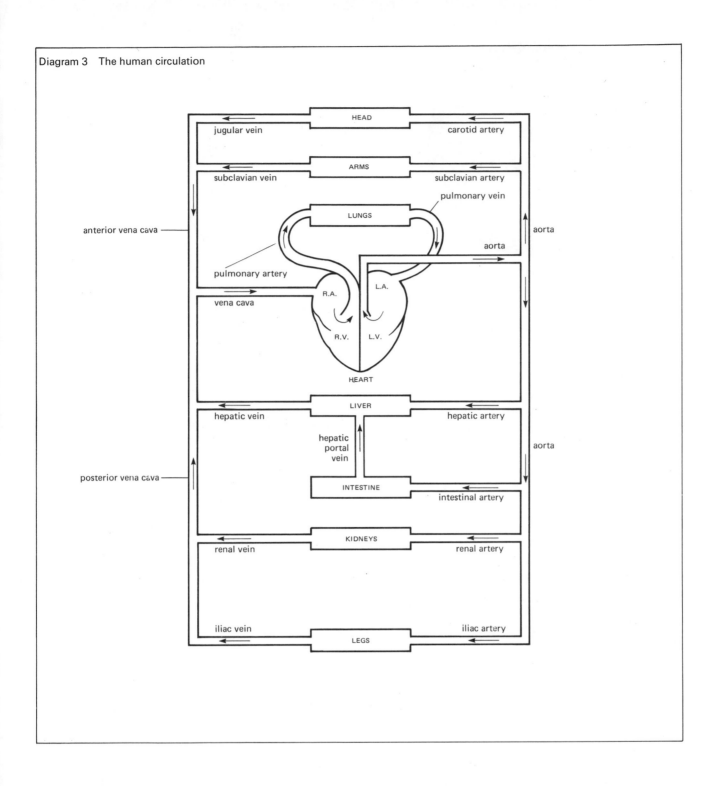

Diagram 3   The human circulation

Blood groups

The human population can be divided into four blood groups known as A, B, AB and O. This is due to the presence or absence of two factors called A and B. If a group A person is given group B blood during a transfusion the bloods do not mix and the person may die. However, group O, which contains neither A nor B factor, can be given to anyone. Group O is the universal donor. Group AB contains both A and B factors so a group AB person can receive any blood. Group AB is the universal recipient. In 1940 it was found that another factor was also involved.

About eighty-five per cent of the human population have this factor and are called rhesus positive, the fifteen per cent who do not have it are rhesus negative. Serious complications can occur if rhesus positive blood, of any group, is given to rhesus negative people as they produce anti-rhesus chemicals.

Related reading

Chapter 18, Structure and function of blood
Chapter 20, The heart

20
The heart

The blood transports many substances around the body:
1 oxygen from the lungs,
2 carbon dioxide to the lungs,
3 food material from the intestine,
4 waste material to the kidneys,
5 HORMONES from the endocrine glands,
6 heat to be distributed,
7 clotting factors to wounds,
8 white cells.

Blood must therefore be pumped around the body. The heart is the pump or, more accurately, two pumps side by side. The right side pumps deoxygenated blood to the lungs, and the left side pumps oxygenated blood to the body. This is called a double circulation. Fish have a single circulation which means that the heart pumps blood through the gills and on to the rest of the body.

In man, the heart is in the THORAX between the lungs. It is surrounded by a delicate MEMBRANE called the pericardium. The pumping action of the heart is driven by cardiac muscles. This is special muscle TISSUE which contracts and relaxes in a natural rhythm without impulses from the nervous system. It also does not become tired like other muscles. This is important because it beats about seventy-five times a minute for the whole of a person's life. The cardiac muscle gets its energy requirements from its own blood supply called the coronary artery and vein, and not directly from the blood in the chambers of the heart (see diagram 1).

Structure of the heart (see diagram 2)

The heart consists of four chambers; the left and right sides are separated from each other by the septum. The upper chambers are called auricles. They have thin walls and receive blood from the veins. It should be noted that another name for auricle is atrium. The lower chambers are called ventricles. They are thick-walled and very

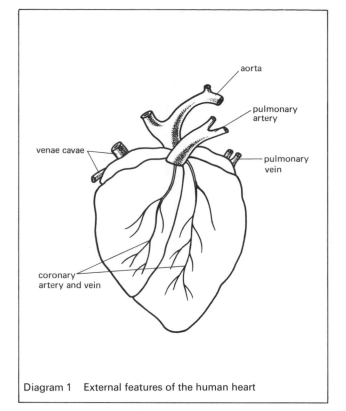

Diagram 1 External features of the human heart

muscular. They receive blood through a valve from the auricles. The walls of the left ventricle are much thicker than those of the right as it pumps the blood all around the body. The right ventricle pumps blood to the lungs. Blood leaves the ventricles through arteries. In the entrance to each artery is a semi-lunar valve which prevents blood flowing back into the heart when the ventricles relax.

Heart cycle

In order to pump the blood, the heart goes through a series of events which repeat themselves about seventy-five times a minute:
1 The heart is relaxed and all the valves are closed. Blood enters both auricles from the veins.
2 The bicuspid and tricuspid valves open and some blood flows into the ventricles.
3 The auricles contract and force their blood into the ventricles.
4 The ventricles contract, closing the bicuspid and tricuspid valves, but the semi-lunar valves are still closed so the pressure on the blood, in the ventricles, rises.
5 The semi-lunar valves open and blood is forced into the arteries.

The whole series of events takes less than one second.

The heart beat is in fact a double-thud, the first being the closing of the cuspid valves when the ventricles contract, followed by the semi-lunar valves closing.

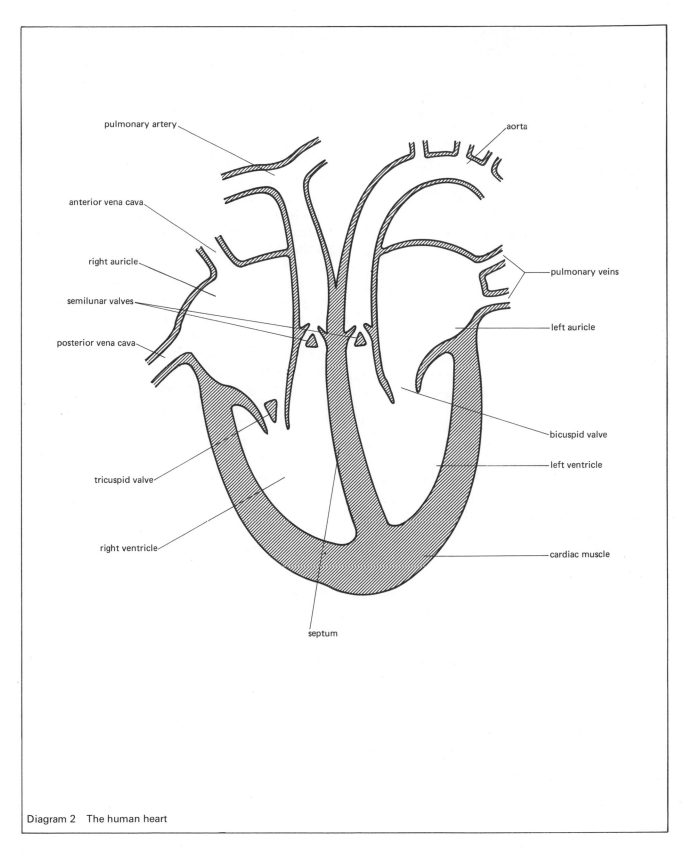

Diagram 2   The human heart

Blood pressure

The spurt of blood from the heart causes the elastic walls of the arteries to stretch and, as they shrink again, they force blood further along. Thus the spurt becomes a continuous flow in the capillaries. The force of this flow is called the blood pressure. Although blood pressure will change during exercise or stress, a very high or very low blood pressure can be dangerous and doctors measure blood pressure to get an indication of someone's health.

Related reading

Chapter 19, Circulation of blood
Chapter 22, Healthy heart

21 Transport in plants

Food from the leaves, water from the soil and oxygen from the atmosphere must reach every living cell in the plant body. Plants have a system of VASCULAR BUNDLES sometimes called veins. These veins contain specialised cells that are like tubes. When a part of a plant, either stem, leaf or root, is cut into a section, the bundles can be seen as a collection of definitely shaped cells (see diagram 2). The bundles are a continuous system of tubes from leaves to root.

A closer examination of one vascular bundle shows two types of tissue, XYLEM and PHLOEM, separated by a band of growing cells called cambium. The vascular bundles are surrounded by larger cells. These large cells have large air spaces between them and are called cortex cells.

The xylem (see diagram 1a)

This tissue is made up of dead cells which are long and strengthened by a woody substance. Water passes from the cells of the root hairs by OSMOSIS into the xylem tubes. Plants are continually losing water by TRANSPIRATION from the leaves, so there is always less moisture in the leafy parts of the plant above ground than in the soil. This difference in the amount of moisture causes water to move by DIFFUSION along the xylem tubes from soil to atmosphere through all the cells of the plant.

The phloem (see diagram 1b)

This tissue is made up of living cells which are long and tube-like. The tubes have perforated ends joining each cell to the next in the vascular system. Phloem cells also have perforated areas on the sides of the tubes allowing substances to pass sideways, as well as up and down the plant. These cells are often called sieve tubes because of the perforated areas. Food which is made in the leaves is passed along these sieve tubes to all parts of the plant and used for growth, activity and storage. The movement of food depends on the phloem cells being living. If the plant is short of oxygen the rate of movement of food is slower. If the phloem cells are killed by heat, for example, no movement of food will take place past those dead cells.

It can be said that water passes from soil to leaves through the xylem by diffusion. Food is carried from the leaves all through the plant by an active process needing living phloem tubes.

Gaseous exchange (see diagram 1c)

Living cells need oxygen to respire. They must also get rid of carbon dioxide waste. These gases are moved by diffusion through the plant in the air spaces of the spongy layer in the leaves, and in the cortex of the stem and root. Gaseous

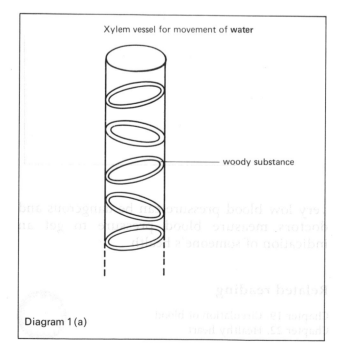

Diagram 1 (a)

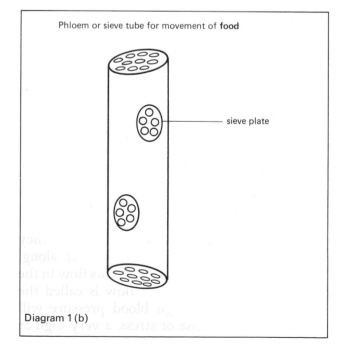

Diagram 1 (b)

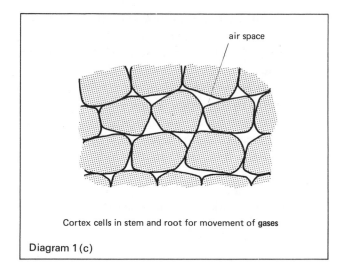

air space

Cortex cells in stem and root for movement of **gases**

Diagram 1 (c)

exchange is complicated by PHOTOSYNTHESIS in the green parts of the plant because the plant takes in carbon dioxide and gives off oxygen. However, respiration continues at all times in the living cells. The concentration of oxygen and carbon dioxide passing into and out of the plant will depend on the amount of photosynthesis taking place.

Related reading

Chapter 6, Photosynthesis
Chapter 7, Transpiration
Chapter 14, Diffusion and osmosis
Chapter 15, Internal respiration
Chapter 25, How plants regulate water

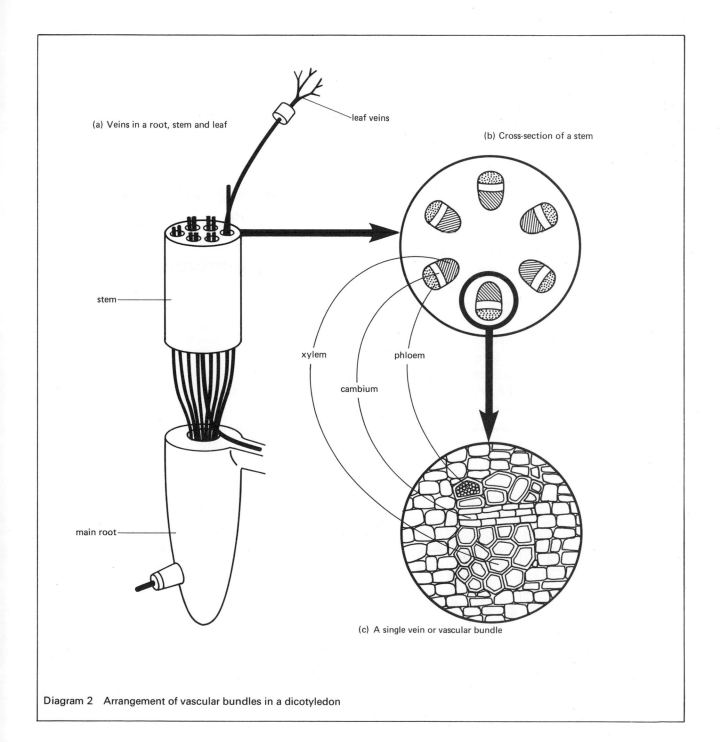

(a) Veins in a root, stem and leaf

leaf veins

(b) Cross-section of a stem

stem

main root

xylem

cambium

phloem

(c) A single vein or vascular bundle

Diagram 2   Arrangement of vascular bundles in a dicotyledon

22
Healthy lungs and heart

Civilised man, living in an industrial ENVIRON-MENT, has a high risk of suffering from certain diseases. The lungs and heart are two ORGANS which are easily damaged by an industrial environment. Severe damage to these organs often causes death.

Lungs

The lining of the lungs is in constant contact with the air. High levels of dust in the air can cause lung disease. Pneumoconiosis is caused by coal dust, and asbestosis is caused by asbestos dust. People who work in dusty air should take special precautions.

The number of people suffering from a bacterial disease of the lungs called tuberculosis (T.B.) has fallen. This decrease is because of injections, mass X-ray and stricter supervision of milk supplies. However, the number of people suffering from lung cancer, bronchitis and emphysema is increasing.

There is a great deal of evidence which links smoking with these diseases (see diagram 3). It has been proved that tobacco smoke seriously affects the covering of small hair-like structures called CILIA, which line the respiratory passages (see diagram 1). These cilia should form a wave action which carries particles of dust and MUCUS away from the lungs. But in a smoker's lungs

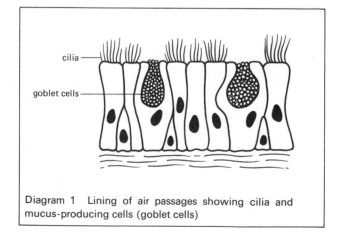

Diagram 1   Lining of air passages showing cilia and mucus-producing cells (goblet cells)

there is a build-up of mucus and dust because the cilia do not work properly. This causes irritation and coughing with an increased risk of infection in the lungs.

Scientists have also shown that sixteen chemicals present in cigarette smoke are capable of forming cancers on the skin of animals. Nicotine causes an increased heart rate and higher blood pressure. Also, because nicotine is an addictive drug, smokers develop the need for a cigarette 'to steady their nerves'. Smoking is an expensive habit. For the sake of your health it is better never to start smoking.

Heart

More than half of all deaths in Britain are caused by atherosclerosis, which is the build-up of fatty, sludgy deposits in the lining of arteries (see diagram 2). This causes the walls of the arteries to thicken and narrow, which can lead to heart disease, strokes and other blood VESSEL diseases.

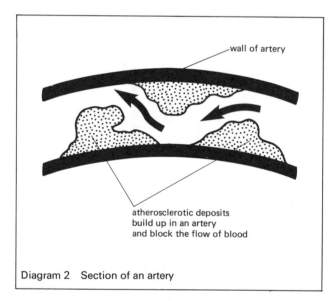

wall of artery

atherosclerotic deposits build up in an artery and block the flow of blood

Diagram 2   Section of an artery

A coronary thrombosis (a heart attack) occurs when a blood clot lodges in the coronary artery and cuts off the blood supply to the heart muscle. Blood clots form easily in people who have a lot of these fatty deposits.

A lot of medical research is being done on the causes of atherosclerosis and heart disease. Many doctors now believe that heart disease is linked to the way we live today. Some research suggests that the main causes of atherosclerosis are (i) smoking, (ii) a diet containing too much fat, (iii) a lot of stress and (iv) a lack of exercise.

Smoking
The risk of coronary heart disease among smokers is twice that of non-smokers and becomes four or five times greater in middle age (35–54 years).

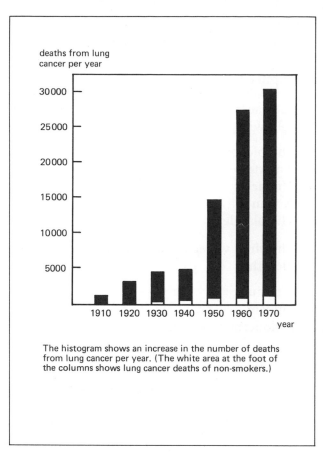

The histogram shows an increase in the number of deaths from lung cancer per year. (The white area at the foot of the columns shows lung cancer deaths of non-smokers.)

Diagram 3 Deaths from lung cancer per year (graph issued by the Cancer Information Association)

No wonder smokers cough.

The tar and discharge that collects in the lungs of an average smoker.

The Health Education Council

A Health Education Council poster

A diet containing too much fat

The level of fat in the diet determines the level of fat in the blood plasma. This should be kept low. Animal fat in butter can be replaced by eating the less harmful fats in some soft margarines. Eating fewer eggs and less cream, but more fish, vegetables and fruit, also helps lower the fat in the blood plasma.

A lot of stress

Everybody is under stress and strain at some time and this can show itself as bad temper, headaches or even diarrhoea. However, stress also has an effect on the heart, causing increased heart rate and blood pressure. People should try to avoid unnecessary stress, perhaps by not driving a car to work. They should also learn methods of relaxation and make sure of a good night's sleep.

Lack of exercise

Everyone should exercise regularly as this helps maintain a good circulation. Swimming exercises every muscle in the body and walking also is excellent. People who have not exercised for years should begin gradually and consult a doctor if there are any unexpected side effects.

High blood pressure

The pressure of the blood in the blood vessels will be higher if the blood vessels are narrower than normal. The narrowing of blood vessels is a natural effect of getting older, but in one in ten people over the age of 45, the increased blood pressure is a threat to health. Blood pressure can be lowered by leading a more restful life, losing weight and not smoking cigarettes. In serious cases, drugs are used to lower the blood pressure.

Related reading

Chapter 17, Breathing in man
Chapter 20, The heart

Questions C

The answers to questions 1 to 7 are shown by one of the letters A, B, C, D or E.

1　Internal respiration is the breakdown of glucose in cells to release:

A　energy and carbon dioxide only.
B　carbon dioxide and water only.
C　energy and carbon dioxide and water.
D　alcohol and carbon dioxide and water.
E　alcohol and water only.

2　The composition by volume of the atmosphere is approximately:

A　one-fifth oxygen and four-fifths carbon dioxide.
B　four-fifths oxygen and one-fifth nitrogen.
C　one-fifth nitrogen and four-fifths carbon dioxide.
D　one-fifth oxygen and four-fifths nitrogen.
E　one-fifth carbon dioxide and four-fifths nitrogen.

3　Oxygen passes from the alveoli into the blood in the lungs by a process called:

A　respiration.
B　breathing.
C　osmosis.
D　diffusion.
E　suction.

4　The main function of haemoglobin in red blood cells is to:

A　kill bacteria.
B　carry oxygen around the body.
C　distribute heat.
D　help the formation of blood clots.
E　carry glucose around the body.

5　Which of the following carries blood to the heart?

A　Vein　　　　　D　Aorta
B　Artery　　　　E　Ventricle
C　Capillary

6　Cardiac muscle can only be found in which of the following places?

A　The biceps
B　An artery
C　The heart
D　A vein
E　The bladder

7　The chief water-conducting tissue of a plant is called the:

A　cambium.
B　phloem.
C　endodermis.
D　exodermis.
E　xylem.

8　Copy out and then complete the following sentences:
(i)　The tissue which forms the supporting rings in the trachea is ....
(ii)　When yeast is used in bread making and brewing the gas given off is...........................
(iii)　The name of the substance used to test for the presence of carbon dioxide is...................
(iv)　The blood vessel leaving the kidney is called.....................................
(v)　The chambers of the heart that receive blood are called................................
(vi)　The food-carrying vessels of a plant are called.....................................
(vii)　The small hair-like structures in the trachea which are paralysed by smoking are called ....

9　Complete each of the following by naming the missing part of the blood system in each series.
(i)　Left ventricle – .... – carotid artery
(ii)　Dorsal aorta – .... – kidney capillaries
(iii)　Intestine capillaries – .... – liver
(iv)　Posterior (inferior) vena cava – .... – right ventricle
(v)　Right ventricle – .... – lung capillaries

10　Write about six lines on FIVE of the following:
(i)　Fermentation　　(vi)　Tidal air
(ii)　Alveoli　　　　(vii)　Gills
(iii)　Capillaries　　(viii)　The cortex
(iv)　Veins　　　　(ix)　The pulse
(v)　Blood pressure　(x)　White blood cells

11 Write an account about twenty-five lines long on ONE of the following. Diagrams may improve your answer.
(i) Smoking
(ii) Functions of blood
(iii) Inhalation (breathing in)
(iv) The xylem
(v) Anaerobic respiration

12 The diagram below shows part of the lung (highly magnified):

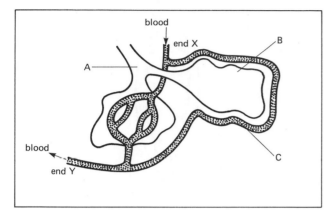

(i) Name the structures indicated by the letters A, B and C.
(ii) In our lungs there are many of the structures labelled B. What is the advantage of this?
(iii) What takes place between B and C?
(iv) Which end, X or Y, of structure C has the highest content of oxygen?
(v) Which end, X or Y, of structure C has the highest content of carbon dioxide?

(YREB)

13

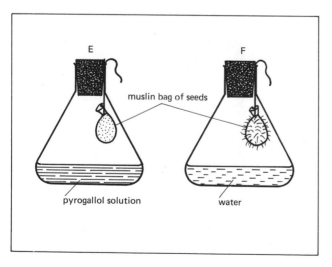

In the experiment represented in the diagram, both E and F had a muslin bag of soaked cress seeds placed inside the flasks one week ago and were left in a warm place.

Explain why the seeds in flask E have not started to germinate.

14 The apparatus below may be used to demonstrate a difference between inspired and expired air.

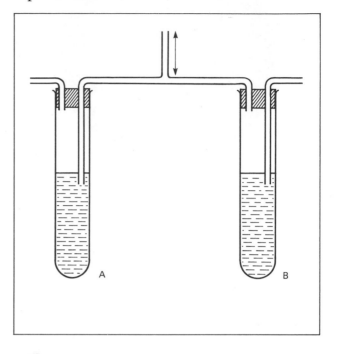

(i) Which is the direction of air flow when the apparatus is being used?
(ii) Name the solution used in tubes A and B.
(iii) After a few breaths what changes would you expect to see in the solution in the tube A and in tube B?
(iv) What conclusions can you draw from this result?
(v) State TWO differences between expired and inspired air which this apparatus does not show.

(Met. REB)

15 (i) List six substances that may be found in blood plasma.
(ii) Describe, with the aid of diagrams, the differences between the structure of arteries, veins and capillaries.

23
The need
for excretion

Excretion is the removal of unwanted substances from the cells of living ORGANISMS. These unwanted substances may be inside the cell for four reasons:

1   Respiration produces the waste substances carbon dioxide and water inside the cell.
2   Unwanted substances are often taken in with food.
3   Occasionally, more food than the cell can use or store is taken into the body.
4   The breakdown of worn-out cells produces unwanted substances.

Excretion is one of the CHARACTERISTICS of all living things because unwanted substances can upset the chemistry of the cell. An example is that too much carbon dioxide will make the CYTOPLASM acidic. Wastes are poisonous, or TOXIC, and can cause the death of particular cells or of the whole organism if the wastes are not removed.

There are four main types of waste substances contained in living cells:

1   Carbon dioxide is produced by respiration.
2   Nitrogenous waste is produced by the breakdown of worn-out cells and by the organism taking in too much protein in the food.
3   EXCESS minerals are occasionally taken into the cells with oxygen and food.
4   Excess water enters the organism from the surroundings by OSMOSIS. This is a problem which affects mainly freshwater animals.

Methods of excretion in animals

Carbon dioxide is excreted as part of external respiration in animals. Nitrogenous waste is removed as harmless chemicals called urea and uric acid.

Simple one-celled animals remove these chemicals, dissolved in water, to the surrounding water by simple DIFFUSION.

The many-celled animals have excretory systems which remove nitrogenous waste, for example:

Diagram 1 shows a flatworm which has a system of canals which branch amongst the body cells. Each branch ends in a special cell called a flame cell. This flame cell draws water and wastes into the canal. This liquid is called urine and is passed out of the body through pores on the surface of the animal.

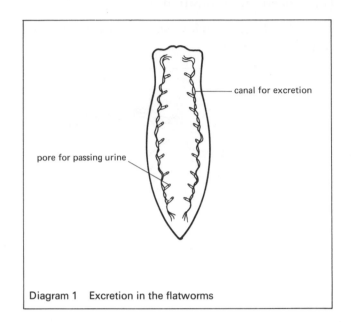

Diagram 1   Excretion in the flatworms

Diagram 2 shows an earthworm which is a segmented worm. Urea is removed from the body cells with water by a system of tubes, in each segment, called nephridia. Urine is passed out of pores on each segment of the body of the worm.

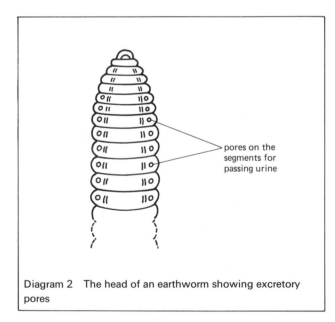

Diagram 2   The head of an earthworm showing excretory pores

Diagram 3 shows the gut of a cockroach. In insects, such as the cockroach, urea is collected from the body by a system of tubes which also reabsorb most of the water in the urine. These tubes are called Malpighian tubules. They empty the urine into the intestine where even more water is reabsorbed.

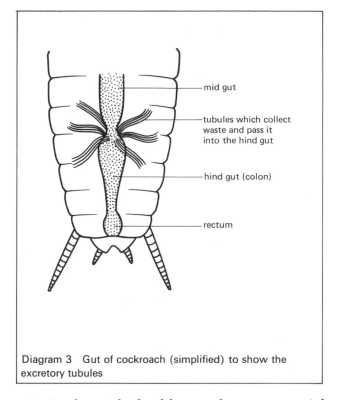

Diagram 3 Gut of cockroach (simplified) to show the excretory tubules

Labels in diagram:
- mid gut
- tubules which collect waste and pass it into the hind gut
- hind gut (colon)
- rectum

Animals with backbones have a special urinary system which includes the kidney as the main organ of excretion of nitrogenous waste.

Methods of excretion in plants

Plants get rid of unwanted carbon dioxide by diffusion through the STOMATA and into the surroundings. Unwanted minerals and nitrogenous substances are changed into insoluble chemicals. These are passed out of the cells and stored in the spaces between the cells. The plant cell wall stops the cell taking in too much water by osmosis so plants do not have excess water to excrete.

Excretion should not be confused with egestion. Egestion is the removal of substances which have not been broken down by digestion. These substances have never been inside the body cells. In man, and in many other animals, the faeces is material which has been taken into the body with food, but has hardly changed at all while in the alimentary canal. Plant cell walls, seeds and mineral substances in the faeces have never been inside the body cells.

Related reading

Chapter 14, Diffusion and osmosis
Chapter 26, Excretion in man

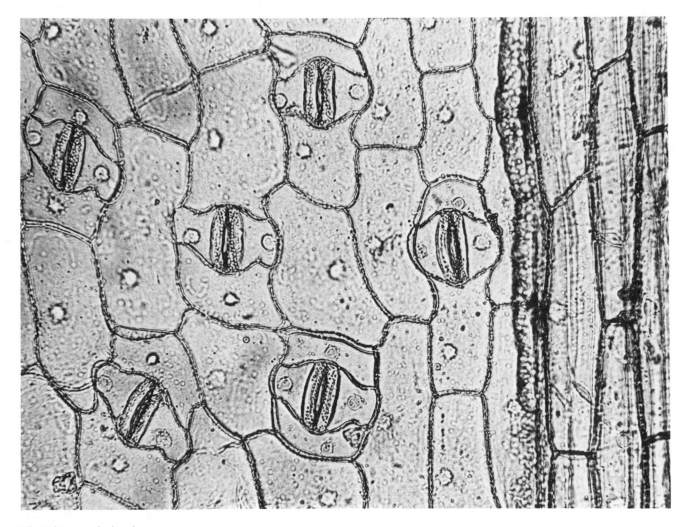

Microphotograph showing stomata

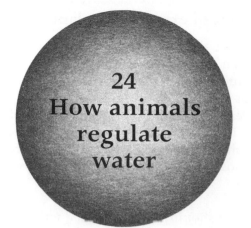

24
How animals regulate water

Living ORGANISMS are made up of cells which are between seventy-five per cent and ninety-five per cent water. Each cell is surrounded by a MEMBRANE. This membrane allows water to pass freely, but regulates the movement of many other chemicals which make up the living cell. Cells must exchange materials with their surroundings. Oxygen and food must pass in and waste products must pass out of the cell. However, the water content must remain reasonably constant.

Water passing into or out of a cell depends on DIFFUSION across the cell membrane. This depends on the concentration of water each side of this membrane. The cell membrane is SEMI-PERMEABLE, therefore this movement of water is called OSMOSIS.

If a cell which is seventy-five per cent water, such as a human red blood cell, is placed in one hundred per cent water, so much water will flow into the cell by osmosis that it bursts (see diagram 1a). This causes a problem for freshwater animals.

If a cell which is seventy-five per cent water is surrounded by a solution with less water, water will flow out of the cell by osmosis (see diagram 1b). This causes a problem for seawater animals.

Controlling the water content of body cells

Animals have developed various ways of controlling the water content of their body cells. This is called OSMOREGULATION. The different ways of control can be divided into two kinds of systems. These are changing the amount of substances dissolved in the cell contents and controlling the amount of water passed out of the body.

Changing the amount of substances dissolved in the cell contents

Some marine animals can change the concentration of their body fluids to match their sea-water surroundings. However too great a difference, like moving them to fresh water, causes the cells to swell and burst.

Controlling the amount of water passed out of the body

Fishes are protected from water passing in and out of the body by their scales, which are resistant to water. However the gills are not protected in this way. Salt-water fishes lose water from the gill cells to the sea because the body cells contain less salt than sea water. To overcome this they take in a great deal of sea water with the food and excrete the salt through special cells in the gill linings.

Freshwater fishes take in water through the gills because their body fluids contain less water and more salts than the fresh water in which they live. To overcome this problem, these fish drink no water. They excrete large quantities of the water as dilute urine.

Amoeba is a tiny one-celled animal which lives in both fresh and sea water. The freshwater type of amoeba takes in water by osmosis. To prevent damage to the animal, the EXCESS water is collected into a space in the cytoplasm called a VACUOLE. When this vacuole reaches a certain size it moves to the edge of the animal and breaks. This gets rid of the excess water (see diagram 2). A sea-water amoeba has no vacuole.

Land animals are in constant danger of losing water to the air. Insects and reptiles have a horny protective outside layer and pass semi-solid urine to conserve water. Birds and mammals also have a waterproof outer layer and pass urine which is high in nitrogenous waste with little water. The gerbil, or desert rat, lives in very dry areas and can remain healthy without ever taking liquid water. The water in the food is sufficient because the kidneys reabsorb water very efficiently. Therefore urine passed by these animals is very concentrated.

Related reading

Chapter 14, Diffusion and osmosis
Chapter 25, How plants regulate water

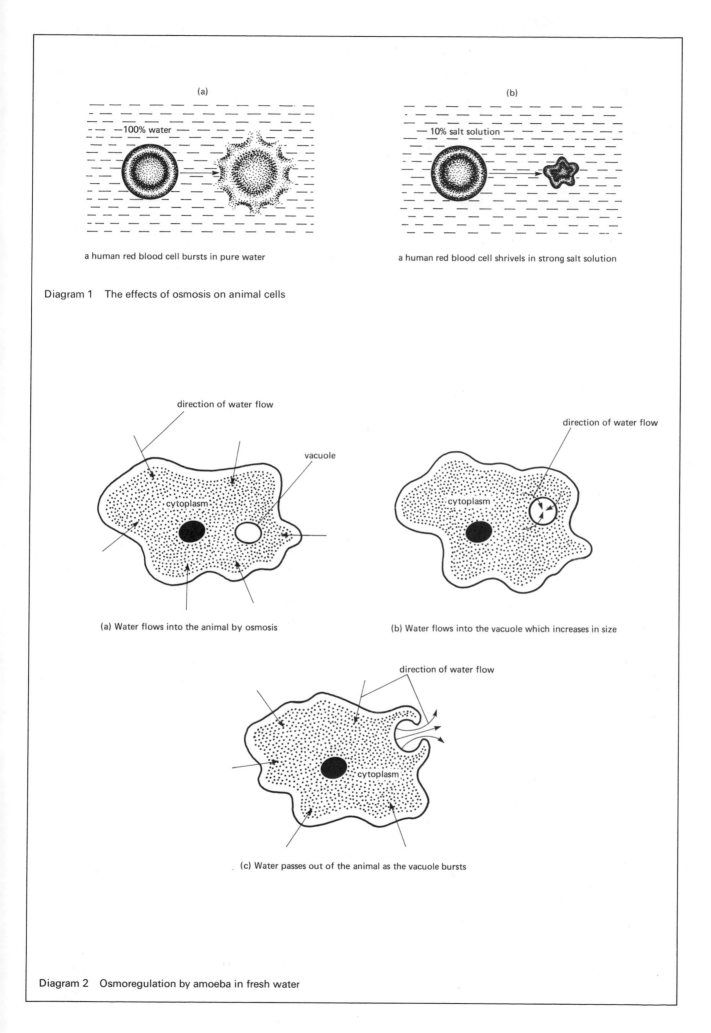

(a)

—100% water —

a human red blood cell bursts in pure water

(b)

— 10% salt solution —

a human red blood cell shrivels in strong salt solution

Diagram 1   The effects of osmosis on animal cells

direction of water flow

vacuole

cytoplasm

(a) Water flows into the animal by osmosis

direction of water flow

cytoplasm

(b) Water flows into the vacuole which increases in size

direction of water flow

cytoplasm

(c) Water passes out of the animal as the vacuole bursts

Diagram 2   Osmoregulation by amoeba in fresh water

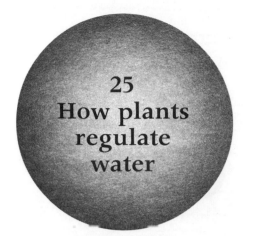

25
How plants regulate water

Plant and animal cells are mostly water. These cells exchange water with their surroundings by DIFFUSION. A cell is surrounded by a MEMBRANE which is SEMI-PERMEABLE. Because of this membrane, diffusion into and out of the cell takes place. This is called OSMOSIS.

The water content of any cell must remain fairly constant if the cell is to continue the many chemical reactions which keep it alive. Plants limit the water taken into the cell in wet conditions, and limit the water lost from the cell in dry conditions.

Control of water passing into the plant cell

Plants which live under water, and the roots of land plants in very wet soil, must not take in too much water. Plant cells have cell walls. This wall allows water to pass freely, but it also acts against the bursting of the cell. The wall pushes against the pressure of osmosis when a plant cell is full of water. This push of the wall prevents too much water flowing into the cell. Too much water would damage the plant cell, but too little water makes the cell soft.

Control of water passing out of the plant cell

Land plants must support their leaves by strong stems so that the leaves are spread in the light and air. For PHOTOSYNTHESIS to take place, the plant needs light and carbon dioxide for the leaves. The support of a plant depends on the cells being full of water. If land plants lose water they WILT and die. The leaves are exposed to a dry atmosphere and are constantly losing water by TRANSPIRATION. This water must be replaced by water passing into the cells from the XYLEM vessels. These xylem vessels receive the water from the root hairs. The root hairs are in contact with the soil and water flows in by osmosis from the soil. If there is plenty of soil water, the cells of the whole plant will be full of water. Then the cells are called turgid (see diagram 1). If soil water is in short supply, for instance during a drought or if a plant has damaged roots, the plant cells become soft and the plant wilts. When plant cells are soft they are called flaccid or plasmolysed (see diagram 2).

In very dry conditions plants often wilt unless they can control the loss of water by transpiration. The openings in the leaves which take in carbon dioxide and lose water are called STOMATA. These stomata have special cells at the openings called guard cells. When these cells are full of starch the plant has enough food stored and photosynthesis can be stopped. The guard cells close and no water is lost. This also means that no carbon dioxide can be taken into the leaf. This is not a very efficient way of preventing water loss as the plant will become short of food. The stomata will open to allow carbon dioxide in so that photosynthesis can start, but water will again be lost.

Summary

Land plants lose water from the leaves by transpiration unless either the stomata are closed or the air around the leaves is very damp. This loss makes the cells of the plant flaccid. This can be changed by water being gained by osmosis from the soil water, through the root hair cells (see diagram 3).

The plant cells become turgid when there is plenty of soil water to be absorbed. Too much water cannot enter the plant cells because the plant cell walls act against the pressure of osmosis and prevent plant cells from bursting.

Related reading

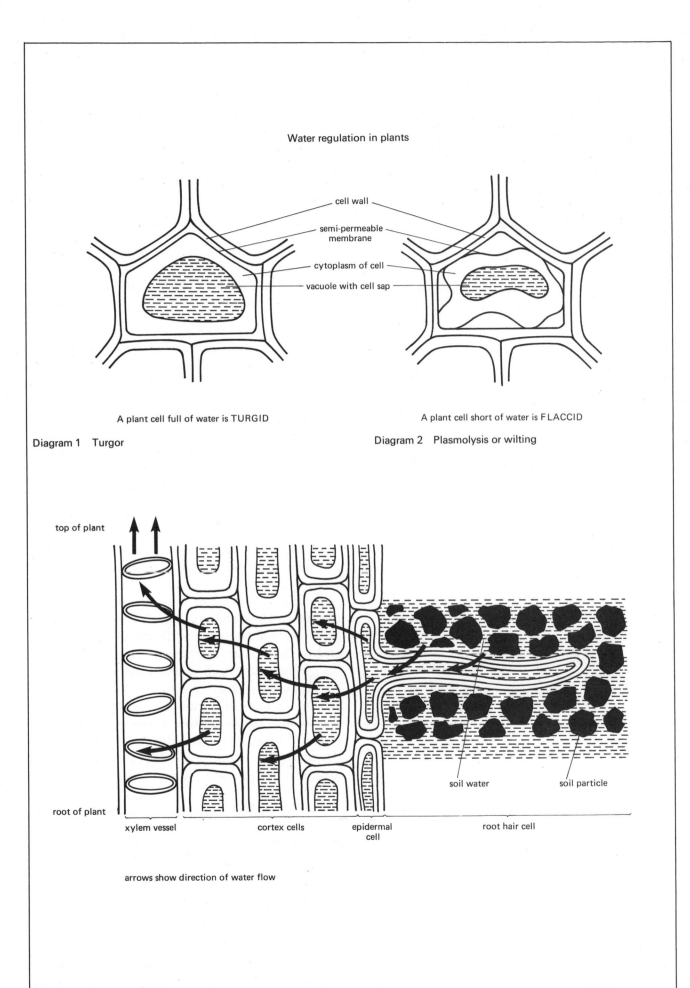

Water regulation in plants

cell wall

semi-permeable membrane

cytoplasm of cell

vacuole with cell sap

A plant cell full of water is TURGID

Diagram 1   Turgor

A plant cell short of water is FLACCID

Diagram 2   Plasmolysis or wilting

top of plant

root of plant

xylem vessel

cortex cells

epidermal cell

root hair cell

soil water

soil particle

arrows show direction of water flow

Diagram 3   Passage of water into a plant

26
Excretion
in man

The chemical reactions in the body which use food to give energy and build new tissue are called METABOLISM. Metabolism produces useful substances but also waste products. These waste products are poisonous and must be removed by the excretory organs. In man the main excretory organs are (i) the kidneys which excrete waste nitrogenous material and help to regulate the amount of water in the body and (ii) the lungs, which excrete carbon dioxide gas and water vapour. The skin excretes some waste products, such as salts and urea, in sweat. The liver excretes BILE PIGMENTS into the ALIMENTARY CANAL. Bile pigments come from the breakdown of worn-out red blood cells. However, the kidneys and lungs are the most important excretory organs.

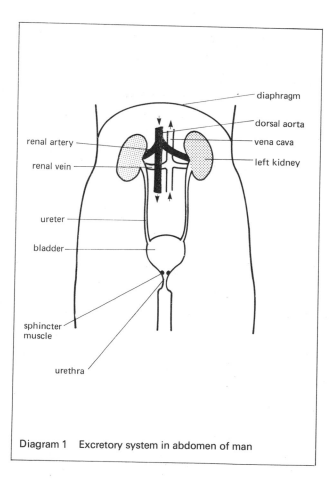

Diagram 1   Excretory system in abdomen of man

The kidneys (see diagram 1)

The kidneys are in the upper part of the ABDOMEN, one on each side of the backbone. Three tubes are connected to each kidney. These are the renal artery, the renal vein and the ureter.

Renal artery
This brings blood containing waste products from the aorta.

Renal vein
This takes blood, with waste products removed, to the vena cava.

Ureter
This carries waste products in urine to the bladder for storage.

Kidney structure (see diagram 2a)
A vertical section of a kidney shows two main zones: the outer ZONE is called the cortex, and the inner zone is called the medulla. The pelvis is where the urine collects before dripping into the ureter. Parts of the medulla project into the pelvis in the form of several cones, called pyramids. Each kidney contains a large number of nephrons in which the work of filtering and water regulation takes place.

Detailed structure and function of a nephron (see diagram 2b)
The renal artery divides into a great many arterioles and CAPILLARIES. These capillaries form a large number of glomeruli in the cortex. Each glomerulus consists of a small knot of a coiled capillary. The capillary entering the glomerulus has a larger diameter than the capillary leaving it, and so there is an increase of blood pressure in the glomerulus. Each glomerulus is surrounded by the swollen end of a kidney tubule called the Bowman's capsule. Pressure forces fluid out of the blood through the walls and into the Bowman's capsule. This filtered fluid contains the nitrogenous waste products in the form of urea, but also glucose, amino acids, salts and a lot of water.

The filtered fluid passes into the first CONVOLUTED TUBULE and the loop of Henlé. In here, much of the water and all the glucose and other useful substances go back into the blood capillaries. This is a complicated process called 'selective absorption'.

In the loop of Henlé and the second convoluted tubule, the amount of water in the urine is regulated. This means that, if a man has lost water through sweating, water is reabsorbed into the blood. The urine in the tubule will be concentrated. If a man has drunk a lot of water,

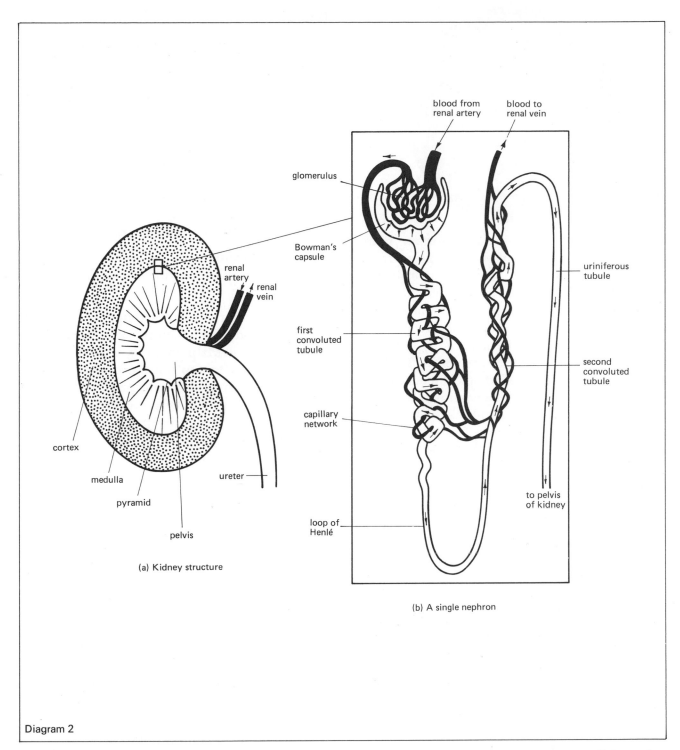

blood from renal artery

blood to renal vein

glomerulus

Bowman's capsule

renal artery

renal vein

uriniferous tubule

first convoluted tubule

second convoluted tubule

capillary network

cortex

medulla

pyramid

pelvis

ureter

loop of Henlé

to pelvis of kidney

(a) Kidney structure

(b) A single nephron

Diagram 2

and the blood is dilute, then more water will remain in the tubule and then be excreted as a more dilute urine. Thus the kidney regulates the amount of water in the blood.

The urine passes along the uriniferous tubule into collecting DUCTS which lead into the pelvis. Urine drips from the pelvis into the ureter and down to the elastic sac called the bladder. Relaxation of the SPHINCTER muscle allows the bladder to empty and the urine to leave the body through the urethra.

The blood passes through the capillaries around the second convoluted tubule and flows into venules which combine to form the renal vein. The blood in the renal vein will have lost almost all its urea and some of its oxygen which was used by the kidney, but it will still contain valuable food and salts as well as the regulated amount of water.

For details about excretion of carbon dioxide and water vapour from the lungs see chapter 17.

Related reading

Chapter 23, The need for excretion
Chapter 17, Excretion of gases from the lungs
Chapter 27, Sweat glands

27
The skin and temperature control in mammals

Mammals have a more-or-less constant body temperature, even though the temperature of their surroundings may change. Mammals and birds are homoiothermic. Homoiothermic means a controlled body temperature, sometimes called warm-blooded. All other animals are poikilothermic. Poikilothermic means a body temperature which varies with the surroundings. These animals are sometimes called cold-blooded.

The mammals must have a way to balance the heat production in their own bodies against the heat lost through the skin. The temperature balance is controlled by the brain and is regulated in the skin by the sweat glands, blood VESSELS and hairs (see diagram).

Sweat glands

When the body starts to overheat, the sweat glands increase their rate of sweat production so that a layer of moisture is produced on the skin surface. Evaporation of the sweat takes heat from the body and so lowers the body temperature.

Sweat consists mainly of water with some salts, urea and lactic acid. In hot climates man may lose up to thirty litres of water and thirty GRAMS of salt in a day. It is important that the salts as well as the water are replaced or heat stroke and cramp may develop.

In some animals, such as dogs, heat is lost from the mouth and lungs by panting.

Blood vessels

The blood CAPILLARIES just beneath the epidermis become wider when the blood is overheated. This causes more blood to flow near the surface of the skin and so more heat escapes to the air. The widening of blood vessels is called vasodilation and it causes the person to appear a flushed pink colour and hot to the touch. Vasoconstriction is the narrowing of the blood vessels near the skin surface, in order to save heat. This makes the person appear pale or faintly blue.

Hairs

The hair erector muscles can contract causing the hairs of the body to stand up. In man this produces 'goose pimples' as we do not have as much hair as some other mammals. Hairs which are standing up trap a layer of still air on the skin surface and this helps to keep in the body warmth. If the body temperature still drops, shivering can begin. Shivering is uncontrolled contractions of muscles which produce heat and raise the body temperature.

FUNCTIONS of the skin in man

Temperature control
This has already been explained in this chapter.

Protection
The surface of the skin is tough and consists mainly of dead cells which are being continually worn away. The Malpighian layer produces new epidermis cells to replace those lost from the surface. The skin is also waterproof and so stops the body drying out. Oily secretions from the sebaceous gland keep the epidermis supple and have antiseptic properties which protect the body against certain bacteria.

Sensitivity
The skin contains various sense organs which enable the body to feel touch, pressure, temperature changes and pain. Certain types of skin receptors are concentrated in particular regions of skin. For example, the finger tips are particularly sensitive to touch due to the large number of touch nerve-endings.

Insulation
The layers beneath the dermis contain fat cells where fat is stored. This fat acts as a heat insulator and helps keep the body warm.

Excretion
Sweat which contains water, urea and salts is lost from the body through the skin.

Vitamin D
The skin can manufacture some vitamin D when it is exposed to sunlight. This is important for healthy growth of bones and teeth.

Related reading

Chapter 23, Excretion
Chapter 32, Skin as a sense organ

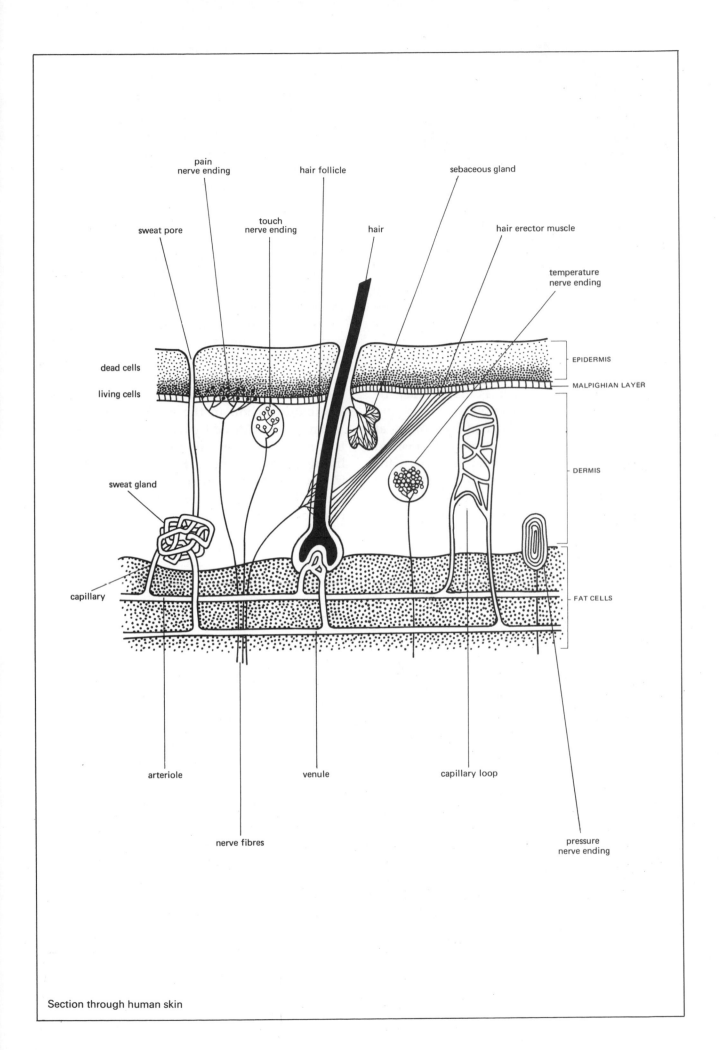

Section through human skin

28
Raw materials

An animal or plant needs energy and raw materials to stay alive, grow and reproduce. The energy needed is provided by the sun. The raw materials needed are provided by the planet earth. The planet only has a certain amount of any substance and there is no way of increasing this fixed amount from any other source. Therefore, anything which is used must be RECYCLED or supplies will run out.

The situation is best explained by looking at just two of the very important minerals needed for living things. Carbon is very important for carrying energy from plants to animals, and nitrogen is very important for body-building in all living things. These elements are carried through the living world and re-used over and over again.

The carbon cycle (see diagram 1)

Animal and plant bodies are about ten per cent carbon. This is a higher percentage of carbon than is in the earth's crust. If carbon were not passed from the living to the non-living world, there would soon be a shortage.

Plants use carbon dioxide from the air and water to make food. Animals eat the food and eventually plants and animals die. The bodies break down because they are eaten by bacteria and fungi. These respire and give out the carbon dioxide back into the air and water, and so the cycle continues.

Some carbon dioxide returns to the air and water when animals give out carbon dioxide as part of respiration.

Plants need the gas carbon dioxide for PHOTOSYNTHESIS, they cannot use pure carbon. Therefore carbon in the earth's crust must be 'burned' or oxidised by respiration with oxygen to turn it into useful carbon dioxide. When living things die, the carbon in their dead bodies is changed to carbon dioxide by bacteria and fungi. These feed on the dead animal and plant remains and release the carbon dioxide.

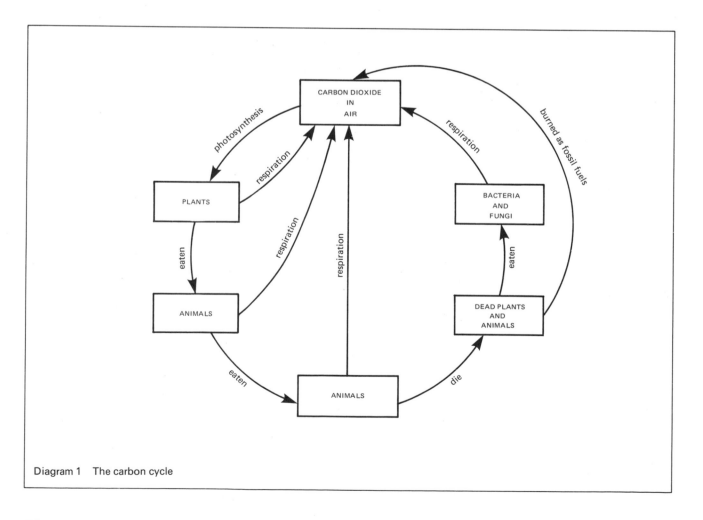

Diagram 1   The carbon cycle

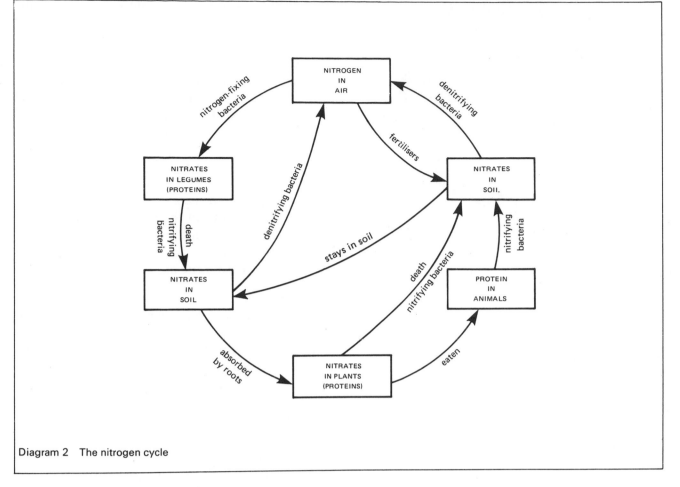

Diagram 2   The nitrogen cycle

The nitrogen cycle (see diagram 2)

Like carbon, nitrogen is present in all living things and there must be a constant supply of a kind that the food-producing plants can use. The air is eighty per cent nitrogen but, like carbon, plants cannot use the pure element. Again bacteria work to combine nitrogen and other substances into nitrates which contain oxygen.

Nitrogen-fixing bacteria

These special and important small ORGANISMS live in small lumps or nodules on certain plant roots. These plants are the pod-producing plants or legumes, e.g. clover, peas, lupins (see diagram 3). These bacteria can change atmospheric nitrogen into nitrates. They make far more than they need to build their own bodies and the EXCESS is used by the leguminous plants to build their proteins. When these plants die, they leave the nitrates in the soil for other plants to use.

Nitrifying bacteria

Animals eat all kinds of plants and so get the vital nitrogen they need to build their protein. When they die, nitrifying bacteria change their decaying bodies back into useful nitrates for more plant growth. Animal waste, such as urine and faeces, is also changed into nitrates by these bacteria.

De-nitrifying bacteria

The nitrogen in the air is kept constant by a third type of bacteria, the de-nitrifying bacteria which change excess nitrates back into atmospheric nitrogen.

Some fertilisers are man-made nitrates and are used to produce extra plant crops with modern farming techniques.

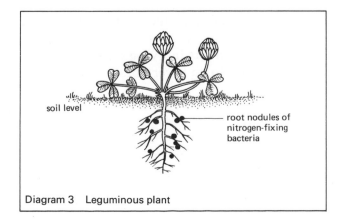

soil level

root nodules of nitrogen-fixing bacteria

Diagram 3   Leguminous plant

Related reading

Chapter 60, Fertilisers
Chapter 62, Conservation of raw materials and fossil fuels

Questions D

The answers to questions 1 to 7 are shown by one of the letters A, B, C, D or E.

1   The structures which are found on the outer layer of roots and which help to absorb water are called:

A   adventitious roots.
B   xylem cells.
C   rhizomes.
D   phloem cells.
E   root hairs.

2   Heat is lost from the body by each of the following processes except:

A   sweating.
B   shivering.
C   urinating.
D   breathing out.
E   defaecation.

3   The structure which carries urine from the kidney to the bladder is called:

A   the colon.
B   the urethra.
C   the pelvis.
D   the oviduct.
E   the ureter.

4   Which of the following is NOT a function of the skin of a mammal?

A   Providing a waterproof layer
B   Manufacturing some vitamin D
C   Controlling the level of hormones
D   Preventing entry of germs
E   Helping to regulate body temperature

5   When a freshwater amoeba is placed in seawater:

A   it loses water.
B   its cell membrane is ruptured.
C   it takes water in.
D   it loses salt.
E   nothing happens.

6   Normal mammalian urine consists of water and:

A   urea.
B   salts and urea.
C   salts, urea and proteins.
D   salts, urea and glucose.
E   salts and proteins.

7   Green plants obtain nitrogen necessary for protein formation in the form of:

A   nitrogen from the air.
B   amino acids from the soil solution.
C   nitrogen from the soil solution.
D   nitrates from the soil solution.
E   ammonia from the air.

8   Copy out and then complete the following sentences:
(i)   Urine is formed in the.............................
(ii)   Water passes into the root hairs of plants by ....
(iii) An animal whose body temperature is variable is said to be..
(iv) The removal of waste products of metabolism from the body is called ....
(v)   Plants use the gas carbon dioxide for ....
(vi) When a red blood cell is placed in distilled water it will ....

9   (i)   State two functions of the roots of a flowering plant.
(ii)   What do nitrogen-fixing bacteria do in the soil?
(iii)   What do nitrifying bacteria do in the soil?
(iv)   What do denitrifying bacteria do in the soil?
(v)   Why should clover roots be ploughed back into soil?
(vi) Construct a diagram to show the nitrogen cycle.

10   Write about six lines on FIVE of the following:
(i)   Waste products in living cells
(ii)   Excretion in an invertebrate
(iii)   Egestion
(iv)   Wilting
(v)   Stomata
(vi)   Sweat glands
(vii)   Hair
(viii) Nitrogen-fixing bacteria
(ix)   Bowman's capsule
(x)   Urine

11 Write an account about twenty-five lines long on ONE of the following:

(i) Temperature control
(ii) The carbon cycle
(iii) Fine structure of the kidney
(iv) Water control in animals
(v) Osmosis

12

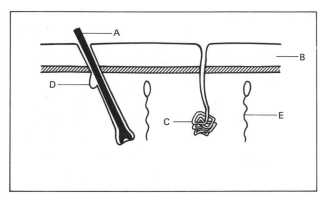

The diagram above is of a section through mammalian skin.

(i) Redraw this diagram and name the parts A–E.
(ii) Give one function of structure A.
(iii) What collects in tube C?
(iv) What is the function of gland D?
(v) Describe the part played by the skin in cooling the body of a mammal.

13

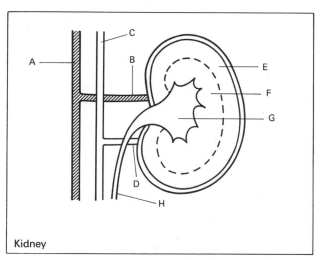

Kidney

(i) Part A is the aorta. Name the other parts, B to H.
(ii) Give one difference between the contents of tube B and tube D.
(iii) Name the fluid passing down tube H.
(iv) Give a simple definition of the term 'excretion'.
(v) Name one other function of the kidney apart from excretion.

14 The table shows the amounts of urine collected from the same group of people in the summer and then in the winter. In both parts of the experiment the diet and activity of the people was the same.

Average volume of urine (cm³)					
	Day 1	Day 2	Day 3	Day 4	Day 5
Summer	990	995	1005	980	1010
Winter	1380	1440	1430	1390	1395

(i) What do you notice about the volumes in winter?
(ii) Carefully explain your answer to (i).
(iii) Why is it essential to keep the diet and activity the same during both experiments?
(iv) What is the main chemical dissolved in urine?

15 (i)

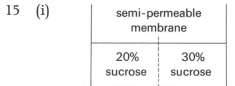

(a) What would happen to the levels of sucrose solution in the above experiment if left for twenty-four hours?
(b) Explain your answer to (a).

(ii)

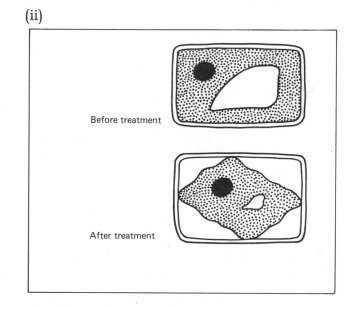

The plant cell above has an osmotic pressure equal to five per cent sucrose. It was placed in a sucrose solution of either two per cent, five per cent or ten per cent.

(a) State one difference you can see in the cell after treatment.
(b) Which of the concentrations of sucrose solution would cause this to happen?
(c) Explain your answers to (ii) (a) and (b).

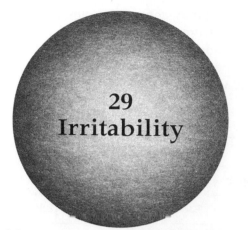

29
Irritability

One of the CHARACTERISTICS of living ORGANISMS is that they respond to their surroundings. This is called irritability. The response usually keeps the organism in a suitable ENVIRONMENT. A man moving into the shade is responding to an over-warm environment.

In order to respond, an organism must receive a STIMULUS of some type from the environment, for example light, heat, touch or chemical substances. Both animals and plants have systems which receive information from the surroundings. This information can lead to behaviour which helps the organism to survive.

How animals respond to stimuli

Animals have a system which is made up of cells that respond to the environment. These cells are called receptors. In this system there are also cells which act on the information received by the receptors to produce an action. These cells are called effectors. When these two kinds of cells are connected by nerve cells they form a nervous system. Receptors and effectors can be connected by chemical substances called HOR-MONES which are part of the endocrine system.

Nervous systems in animals vary from those that are simple to those that are very complicated.

Hydra

A simple nervous system in animals can be found in the small freshwater animal called Hydra. This animal has a nerve net of simple nerve cells (see diagram 1). This nerve net makes it possible for Hydra to respond only as a whole animal to a stimulus on one part of the body.

Earthworms

Earthworms have a centre to the simple nervous system which makes it possible for the animal to control its responses to the environment. A nerve cord runs down the underside of the worm and nerves run out into each segment. (See diagram 2.)

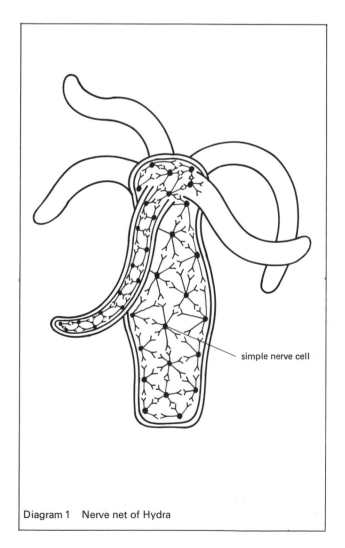

Diagram 1 Nerve net of Hydra

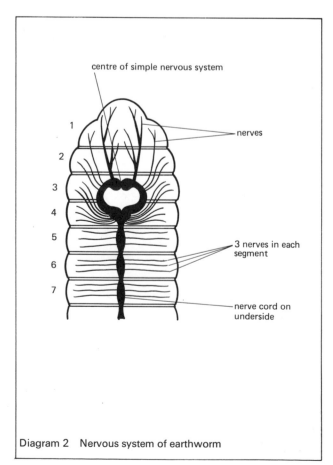

Diagram 2 Nervous system of earthworm

Insects

Insects have a collection of nerve TISSUE which represents a brain. This brain is in the head and is continuous with a nerve cord which runs down the underside of the insect. The insect also has special sense organs for sight. These are compound eyes. Other stimuli are sensed by small pointed projections all over the body. These projections are very plentiful in the antennae. (See diagram 3.)

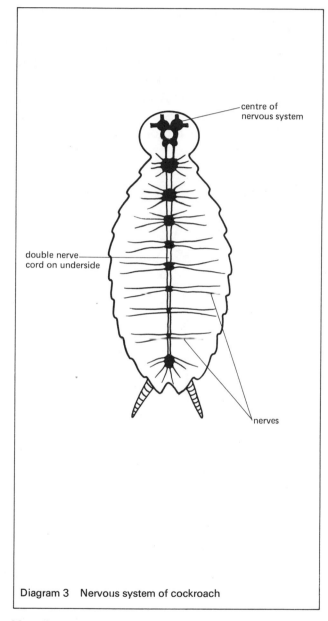

Diagram 3   Nervous system of cockroach

Vertebrates

Vertebrates, such as the mammals, have a complex nervous system and brain. These animals also have a complex hormonal system. These are discussed in chapters 30, 31 and 35.

All animals have glands or cells which produce chemical substances called hormones. These hormones cause the animal to respond to stimuli more slowly than nerves.

How plants respond to stimuli

Plants have only hormones to connect the receptor cells, which receive stimuli, and the effector cells, which respond to the stimuli. The range of plant responses is very similar to the range of animal responses but, because there is no nervous system, plants respond more slowly than animals to their environment. Plant responses are made possible mostly by growth. A plant bending towards light or towards the support of a wall does so by growing. Certain plants, however, respond to touch more quickly than growth could make possible. Mimosa leaflets, when touched, collapse quickly due to loss of water. The Venus fly-trap (see diagram 4) closes quickly due to water increasing in certain cells. This response is triggered by hairs on the trap, but the way in which the stimulus is carried to the cells which increase their water content is not known.

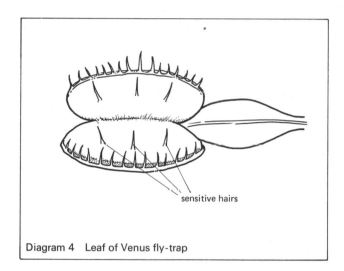

Diagram 4   Leaf of Venus fly-trap

Plants have no nerve cells, nerve tissues or nervous system, therefore hormones are the only means by which the message from receptors to effectors is carried in plant tissue.

Related reading

30 Nerve messages

The nervous TISSUE in animals is made up of special cells called neurones. Neurones conduct the nerve messages around the body.

Neurones

The vertebrate nervous system has three types of neurones: sensory neurones, intermediate neurones and motor neurones.

Sensory neurones (see diagram 1) are composed of CYTOPLASM surrounded by a cell MEMBRANE and containing a NUCLEUS. Each sensory neurone has a single strand of cytoplasm called a dendron. This dendron carries the nerve message from a receptor, such as a pain receptor in the skin, to the cell body. A second strand of cytoplasm called the axon carries the nerve message away from the cell body to the next neurone.

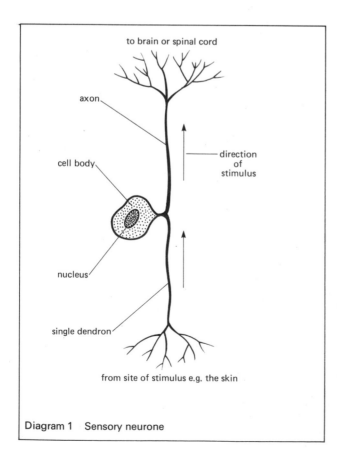

Diagram 1   Sensory neurone

Intermediate neurones (see diagram 2) are found in the spinal cord or in the brain. Intermediate neurones have many short strands of cytoplasm which carry nerve messages to other neurones.

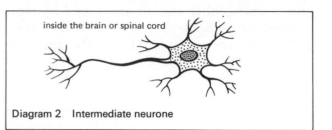

Diagram 2   Intermediate neurone

Motor neurones (see diagram 3) are surrounded by strands of cytoplasm called dendrites. The dendrites pick up the message from intermediate neurones. An action in a muscle is the result of a nerve message which passes from the cell body of the motor neurone, along the axon, to the muscle.

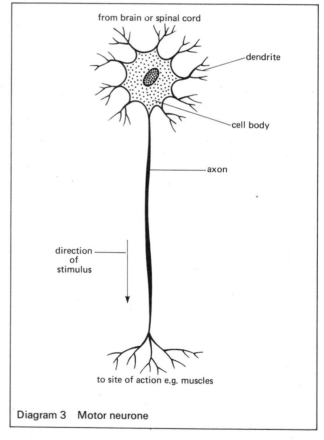

Diagram 3   Motor neurone

Nerve messages

A nerve message is an electrical impulse which travels very quickly along the axon or dendron of a neurone.

The synapse

Neurone endings do not touch inside the nervous system. Electrical impulses are changed into chemical substances which cross the gap, called a synapse, between different neurones.

Nerves

The axons and dendrons of neurones connect every part of the body to the brain and spinal cord. These strands of cytoplasm are bound together to form nerves. Each axon or dendron is insulated from all others. The axon or dendron plus insulation is called a fibre. Many fibres bound together make up a nerve. Most nerves are mixed, which means they carry both sensory and motor fibres. Some nerves are entirely motor or entirely sensory.

Most nerves are not linked directly to the brain but to the spinal cord where the intermediate neurones pass the message on to the brain.

Intelligent actions

In humans, most reactions to outside conditions are under the conscious control of the individual. When a hot dish is picked up and pain is felt, a human being can decide to hold on to the dish, although he would probably drop a worthless object. This type of reaction to a stimulus from the surroundings is intelligent and involves the brain.

Reflex actions (see diagram 4)

Certain actions of humans cannot be controlled by the conscious mind. This is because the path of the nerve message does not include the brain.

These actions are called reflex actions. The knee jerk and the reaction of the pupil to bright light are such actions.

When the tendon below the knee is given a sharp tap, a receptor picks up the STIMULUS and passes a nerve message along the sensory fibre to the cell body of the sensory neurone. The message continues along the axon of the sensory neurone into the spinal cord, across a synapse to an intermediate neurone. The intermediate neurone passes the message across another synapse to the dendrites of a motor neurone in the spinal cord. The message then continues along the axon of the motor neurone to the muscle above the knee. This muscle is stimulated to contract and the leg kicks up.

Learned behaviour

Very few human reactions do not involve the brain. Human behaviour is learned by imitation and by trial and error. Conscious control of reactions is highly developed in all mammals, and most highly developed in man.

Related reading

Chapter 31, Central nervous system

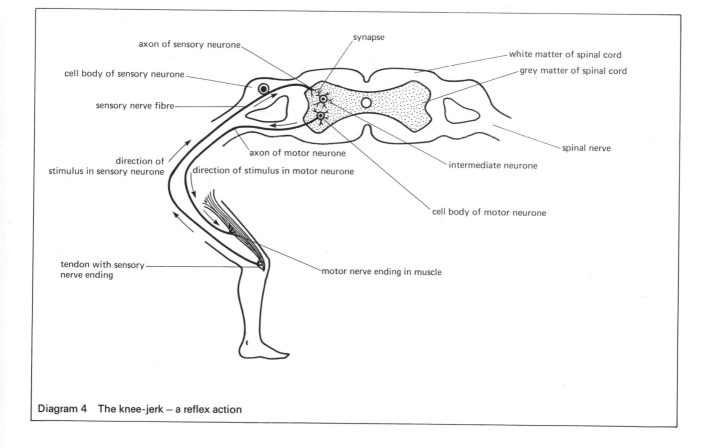

Diagram 4   The knee-jerk — a reflex action

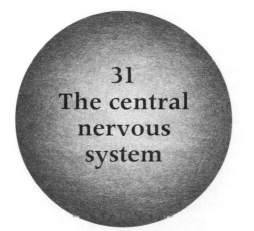

31
The central nervous system

The vertebrates are animals that have a backbone. There is a nerve cord inside the backbone. From this cord, nerves run to and from every part of the body. At the head-end of the nerve cord, also called the spinal cord, there is a mass of nervous cells which is called the brain. The brain and spinal cord form the central nervous system of vertebrates.

The central nervous system of the human

The spinal cord

The spinal cord consists of a tube made up of three layers (see diagram 1):

(i)   The central canal is filled with a fluid similar to LYMPH and called the cerebro-spinal fluid.

(ii)   The grey matter is made up of the bodies of the nerve cells or neurones and their connections with each other.

(iii)   The white matter is made of nerve fibres. These fibres run lengthways and link the different parts of the cord and the brain. Each fibre is insulated from other fibres by a substance called myelin. Myelin gives this layer its white colour.

The functions of the spinal cord are mainly to link the nerves of the body together and to link the nerves with the brain (see diagram 2).

The brain

The human brain has a similar pattern to that of all other vertebrates. However, certain parts of the human brain are more highly developed. The brain is divided into a number of regions (see diagram 3):

(i) **The medulla oblongata**   This region is where the spinal cord joins the brain. This part controls the unconscious actions of the body like breathing, heart beat and movement of food along the gut.

(ii) **The cerebellum**   This is the part of the brain which controls the way muscles act together for any particular activity such as writing or walking. This part of the brain keeps muscles working together to enable humans to sit or balance without having to think consciously about these activities.

(iii) **The cerebral hemispheres or cerebrum**   These are the largest parts of the human brain and they overlap the whole of the front and sides. In man the surface is deeply folded to increase the SURFACE AREA for nerve cells. Different parts of the hemispheres are linked together by nerve fibres inside the grey matter. The cerebral hemispheres, or cerebrum, are the areas of the brain which enable man to be conscious of his actions. The many linkages between neurones provide man with intelligence and memory.

(iv) **The ventricles**   The ventricles of the brain are similar to the central canal of the spinal cord and also contain cerebro-spinal fluid.

(v) **The pituitary gland**   This ductless gland, or endocrine gland, is not involved in nerve messages but is involved in chemical messages. These messages are sent by the secretion of chemicals, called HORMONES, into the blood system.

(vi) **The meninges**   These are three layers of TISSUE which surround the whole of the brain and spinal cord:

The outer layer is of tough tissue which separates the nervous tissue from the bone.

The middle layer is spongy tissue which acts as a shock-absorber for the nerve cells.

The inner layer is a thin layer in very close contact with the nerve cells. This layer carries blood to the nerve cells making the exchange of waste and the supply of food and oxygen possible.

Mapping brain functions (see diagram 4)

Medical science has been able to map the brain of humans because man can describe what he feels when parts of the brain are stimulated artificially under local anaesthetics in hospital conditions.

Related reading

Chapter 30, Nerve messages
Chapter 35, Pituitary gland

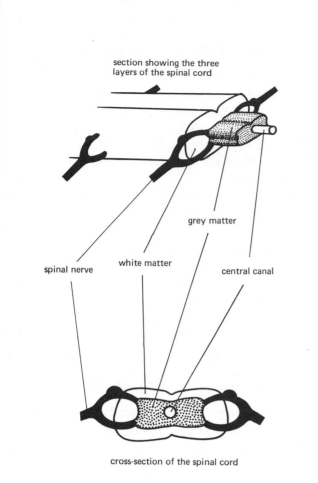

section showing the three layers of the spinal cord

grey matter

spinal nerve

white matter

central canal

cross-section of the spinal cord

Diagram 1   The spinal cord

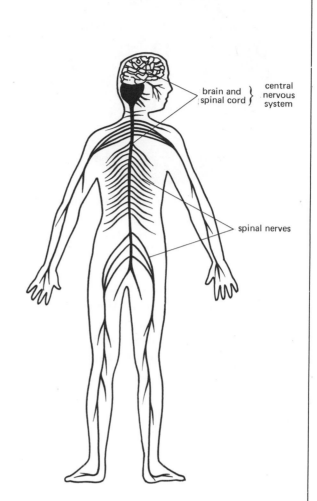

brain and spinal cord } central nervous system

spinal nerves

Diagram 2   Parts of the nervous system

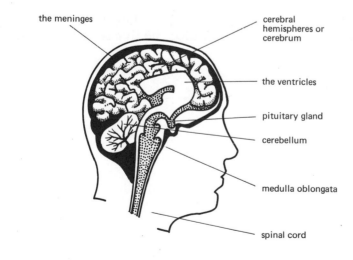

the meninges

cerebral hemispheres or cerebrum

the ventricles

pituitary gland

cerebellum

medulla oblongata

spinal cord

Diagram 3   The parts of the brain

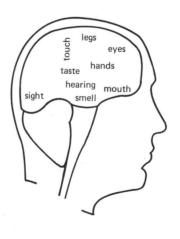

legs

touch

eyes

hands

taste

hearing

mouth

sight

smell

Diagram 4   A map of brain functions

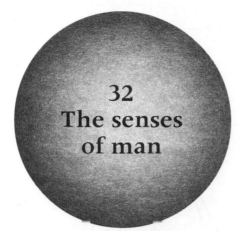

32
The senses of man

The brain must be able to collect information about the surroundings so that it gives the correct instructions to the body. Information is collected by special receptors called sense cells and this information is transmitted to the brain by nerves. The information is called a STIMULUS and the sense cells convert the energy of a stimulus into electrical energy.

In humans there are five main parts of the body which receive various stimuli:

Skin which receives touch, pressure, temperature and pain.
Nose which receives smell.
Tongue which receives taste.
Ears which receive sound (chapter 33).
Eyes which receive light (chapter 34).

Sense cells are also found in the internal organs such as the gut. These inform the brain about internal conditions such as hunger or tiredness.

Senses of the skin (see diagram 1)

The skin contains millions of separate tiny sense organs of different types.

The sense of touch allows man to feel different textures and fine detail. The touch receptors are spread all over the body, but they are very close together in the finger tips. Some blind people can read 'Braille', which is a system of raised dots on paper which they feel with their finger tips. The sense of pressure allows man to hold objects without crushing them and to use delicate instruments. The pressure receptors are usually deeper down in the skin than those for touch. Pressure receptors can also be found in the muscles and joints. Pressure receptors are necessary so that the brain knows the positions of the limbs and the state of the muscles.

The sense of temperature depends on two types of receptors, one for hot and the other for cold. However, the receptors cannot tell an exact temperature. If you are cold, a warm bath feels hot, but, if you are hot, the same warm bath feels cold.

The sense of pain is important as it warns the brain that something is wrong with the body. Pain receptors can be found throughout the body and not just in the skin.

Sense of smell (see diagram 2)

Smell is important in many animals for finding food, avoiding enemies and recognising friends. Man can use his sense of smell for protection, for example, he can smell smoke or food which has decayed. However, he can also get pleasure from certain smells such as the smells of good cooking or perfume.

The small receptors, also called olfactory organs, are high in the nasal cavities. They are stimulated by chemicals which become dissolved in the moisture within the nose.

The sense of smell is reduced when man suffers from 'flu or a cold. This is because MUCUS covers the smell receptors. The sense of taste also appears to be reduced during a cold, which means that many 'tastes' are in fact a mixture of taste and smell.

Sense of taste (see diagram 3)

Taste receptors are stimulated by chemicals in solution. They are grouped together into little projections called taste buds which are situated in certain areas of the tongue. There are four types of taste bud, each sensitive to a different taste. They are sweet, sour, salt and bitter. The taste of different foods depends on how much each of the four types of taste bud are stimulated.

Touch, taste and smell, along with hearing and sight, are often called the five senses of man.

Related reading

Chapter 33, The ear
Chapter 34, The eye
Chapter 27, The skin

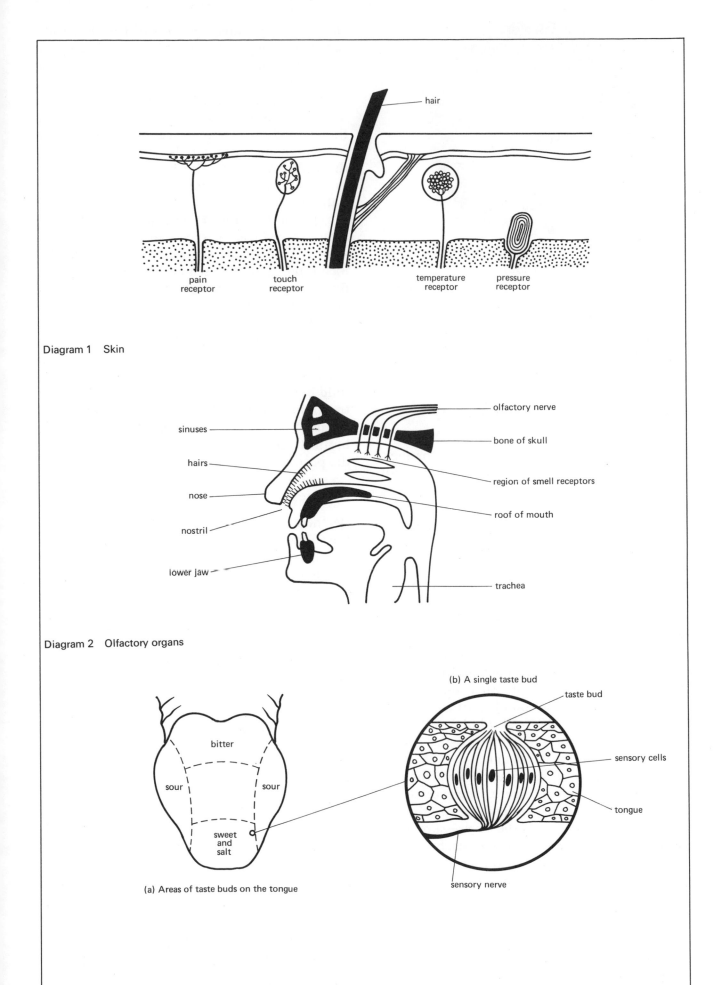

hair

pain receptor

touch receptor

temperature receptor

pressure receptor

Diagram 1   Skin

olfactory nerve

sinuses

bone of skull

hairs

region of smell receptors

nose

roof of mouth

nostril

lower jaw

trachea

Diagram 2   Olfactory organs

(b) A single taste bud

taste bud

bitter

sensory cells

sour

sour

tongue

sweet
and
salt

sensory nerve

(a) Areas of taste buds on the tongue

Diagram 3

33
The ear

Sound is produced by an object vibrating in the air. The vibrations form sound waves in the air. The sound waves travel in a similar way to the ripples made when a stone is thrown into a pool of water. The ear can receive these sound waves and convert them into nervous impulses which are sent to the brain. The brain can change these impulses into signals which have meaning.

Structure of the ear (see diagram)

Outer ear
The pinna collects the sound waves and channels them into the auditory canal. The auditory canal contains wax-SECRETING glands and small hairs which prevent the entry of dust or insects. At the end of the auditory canal is the ear drum (also called the tympanic MEMBRANE).

Middle ear
The sound waves strike the ear drum and cause it to vibrate. This vibration is carried and amplified by the auditory ossicles. The ossicles are three bones called the hammer (or malleus), the anvil (or incus) and the stirrup (or stapes). These are the smallest bones in the body and are held in place by muscles. When the ear drum vibrates the ossicles amplify the sound. This means that the stirrup vibrates the oval window at about twenty times the original force of the sound waves received at the ear drum.

The middle ear is filled with air which is usually at atmospheric pressure. However, when the air pressure changes in the outer ear, the ear drum would be damaged if the air pressure were not the same on both sides. The eustachian tube connects the middle ear to the throat. This allows air to enter or leave the middle ear and so balances the pressure in the middle ear with the pressure of the atmosphere outside.

Inner ear
The inner ear has two main parts. The function of one part is hearing and the function of the other is balance.

Hearing   The inner ear is filled with a fluid called perilymph. The fluid is set vibrating by the movement of the oval window and these vibrations are passed to the cochlea. The cochlea is a tube that rises in a spiral and contains many hair-like sensory cells. These are STIMULATED by the amplified sound waves carried by the perilymph and send impulses to the brain along the auditory nerve. The brain then interprets these impulses as sounds.

Balance   Humans can tell if their bodies are upright or bending to one side, and they are aware of changes of speed and direction. This sensitivity depends on many different stimuli, including sight. The semi-circular canals in the inner ear detect changes in the direction of movement or balance even when the eyes are closed.

The semi-circular canals consist of three fluid-filled tubes arranged at right angles to each other. When the head moves to one side, the fluid in the canals also moves. This movement of the fluid is detected by nerve endings in the canals. The nerves then send impulses to the brain from which the brain can interpret the position of the body.

Deafness

Deafness may be caused by a blockage in either the auditory canal or the eustachian tube. It may also be caused by damage to the ear drum or ossicles. All of these prevent transmission of sound vibrations, but their effects can sometimes be lessened by hearing aids.

Deafness caused by failure of the auditory nerve, or damage to the hearing centres of the brain, cannot be helped by hearing aids.

Related reading

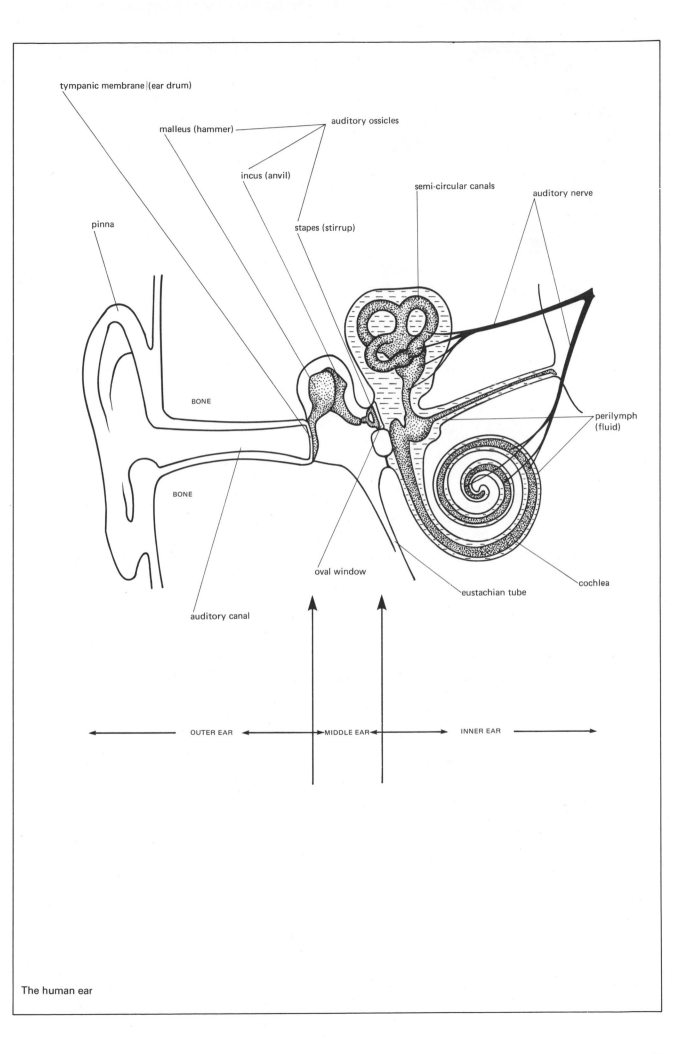

The human ear

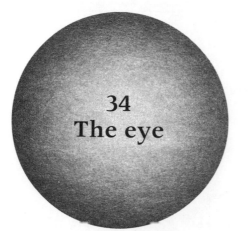

34
The eye

Many animals are sensitive to light, but only the vertebrates, some arthropods and some molluscs are able to form picture images of the outside world. The organ of sight is the eye and its STRUCTURE is similar in all vertebrates.

Structure of the human eye (see diagram 1)

Eye lids
These can cover and protect the eye. Fluid is spread over the surface of the eye during blinking and this prevents the eye from drying.

Eye muscles
The eyeball is held by six muscles in a socket, called the orbit. These muscles move the eyeball so that the eyes can follow moving objects and look together in a chosen direction.

Conjunctiva
This is a thin transparent layer which lines the inside of the eyelids and continues across the front of the eyeball.

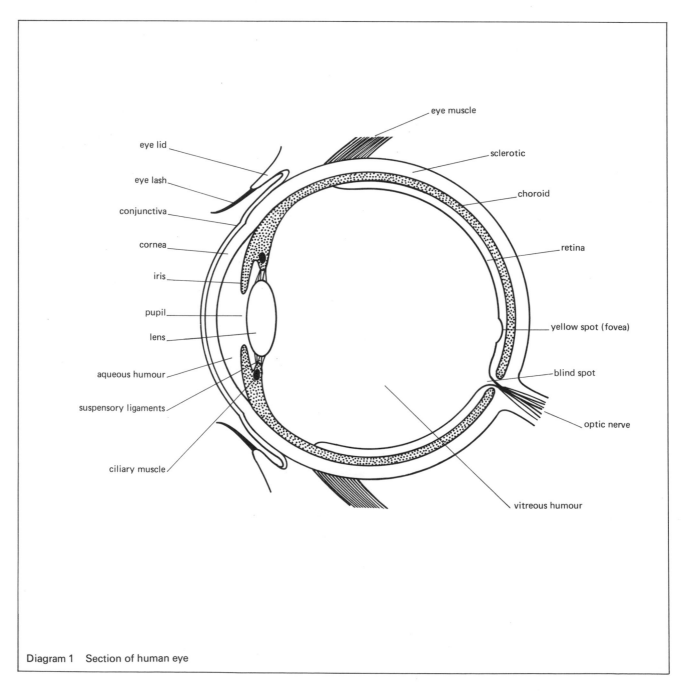

Diagram 1   Section of human eye

Cornea

The cornea is the transparent front part of the sclerotic. The cornea bends the light rays as they enter the eye so that, after passing through the lens, they can be focused on the retina.

Iris and pupil

The iris is the coloured part of the eye (see diagram 2). It has a hole in the middle called the pupil. The pupil appears black because the inside of the eye is dark. The iris controls the amount of light entering the eye, by changing the size of the pupil. When a bright light shines on the iris, the pupil becomes smaller and so allows less of this bright light into the eye.

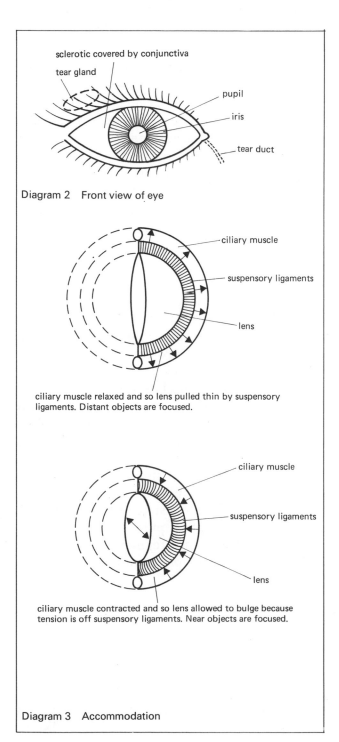

Diagram 2   Front view of eye

(labels: sclerotic covered by conjunctiva; tear gland; pupil; iris; tear duct)

(labels: ciliary muscle; suspensory ligaments; lens)
ciliary muscle relaxed and so lens pulled thin by suspensory ligaments. Distant objects are focused.

(labels: ciliary muscle; suspensory ligaments; lens)
ciliary muscle contracted and so lens allowed to bulge because tension is off suspensory ligaments. Near objects are focused.

Diagram 3   Accommodation

Lens

The lens continues the bending of light, called refraction, started at the cornea, so that an image is formed on the retina. The shape of the lens can change to focus near or distant objects. This is called accommodation (see diagram 3).

Aqueous and vitreous humours

The aqueous humour is a watery solution in between the cornea and the lens. It carries oxygen and food to the cornea and lens which have no blood vessels.

The vitreous humour is jelly-like and fills the space between the lens and the retina. Both of the humours help to keep the shape of the eyeball.

Sclerotic

This is the tough, white outside layer of the eye. It helps maintain the shape of the eye.

Choroid

The choroid is the layer of tissue lining the inside of the sclerotic. The choroid contains blood vessels which carry food and oxygen to the retina.

Retina

The inner layer at the back of the eye is called the retina. It contains two types of cells which are sensitive to light. These cells are called the rods and the cones. The rods are sensitive to low intensity light and allow the eye to see black, white and shades of grey. The cones are sensitive to brighter light and allow the eyes to see colours.

Fovea

The yellow spot, or fovea, is a point on the retina where there is a high concentration of cones but no rods. When you look directly at an object, the image falls on the fovea.

Optic nerve and blind spot

The light-sensitive cells of the retina transmit their image to the brain through the optic nerve. At the point where the optic nerve passes through the retina there are no light-sensitive cells and so it is called the blind spot.

Stereoscopic vision

Each eye forms its own image so that two messages are sent to the brain. The brain combines these two images so that we can judge distance and see the 'three dimensions' of an object.

Related reading

Chapter 32, Taste, touch, smell
Chapter 33, The ear

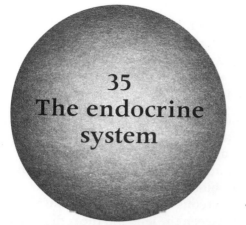

35
The endocrine system

In the animal kingdom, tadpoles grow into frogs and caterpillars into butterflies. In humans, the young grow up and change into adults.

These changes, and many others, must be organised and CO-ORDINATED. This co-ordination is carried out by chemical messengers called HORMONES.

Endocrine system in man

Hormones in humans are made in special glands called endocrine or ductless glands. They are named 'ductless' because their cells pass the hormones straight into the blood, rather than into a pipe or DUCT (see diagram 1).

Hormones are therefore carried all over the body by the blood and do their job in many different parts of the body, away from the gland where they were first made. Hormone messages are slower to act than nerve messages as they travel at the speed of the blood. Most of them co-ordinate the long-term activities of the body such as growth, development and METABOLISM.

There are five main endocrine glands (see diagram 2). These are the pituitary gland, the thyroid gland, the pancreas, the adrenal glands and the gonads.

The pituitary gland
The pituitary gland is on the underside of the brain. It SECRETES large numbers of hormones. Some of these affect the body directly, such as the growth hormones and those which affect blood pressure.

However, many of the hormones control the other endocrine glands so the pituitary is sometimes called the 'master ductless gland'.

The thyroid gland
The thyroid gland is made up of two lumps on either side of the trachea (windpipe) in the neck. It secretes the hormone called thyroxine which contains iodine. A diet lacking in iodine can lead to thyroid inactivity which includes a general slowing down of the rate of metabolism. In the young this means that both physical and mental growth will be retarded. Adults who suffer from thyroid inactivity appear lazy and dull.

Over-activity of the thyroid causes too fast growth in young, and adults who are nervous, active and excitable.

The pancreas
The pancreas contains ducted glands which secrete digestive juices. It also contains ductless glands, called 'Islets of Langerhans', which secrete a hormone called insulin. Insulin regulates the amount of sugar in the blood. Too little insulin causes sugar diabetes. This occurs when the blood sugar level is too high and therefore sugar is excreted in the urine. Diabetes can be controlled by regular injections of insulin.

The adrenal glands
The adrenal glands are just above the kidneys. They have two regions. The outer part, called the cortex, secretes hormones called steroids which affect sodium in the body. The inner part, called the medulla, secretes a hormone called adrenalin. Adrenalin is secreted when the body senses danger or stress. It causes an increase in pulse rate and diverts blood from the ALIMENTARY CANAL and skin to the muscles.

The gonads (reproductive organs)
The gonads are the reproductive organs which are the testes in the male and the ovaries in the female.

In the male, hormones secreted from the testes stimulate the growth of hair on the face, chest and legs, the deepening of the voice and the other changes of a boy growing into a man. They also affect sexual behaviour.

In the female, the hormones control the changes from girl to woman, such as the development of breasts, widening of the hips and the growth of pubic hair. One of the hormones, progesterone, prepares the woman's body for pregnancy and stops the development of more ova (eggs) while the baby is growing inside the mother. Progesterone is used in some types of 'contraceptive pill'. A woman taking this pill cannot become pregnant as the hormone is stopping the development of ova.

Related reading
Chapter 29, Response to stimuli
Chapter 36, Plant hormones

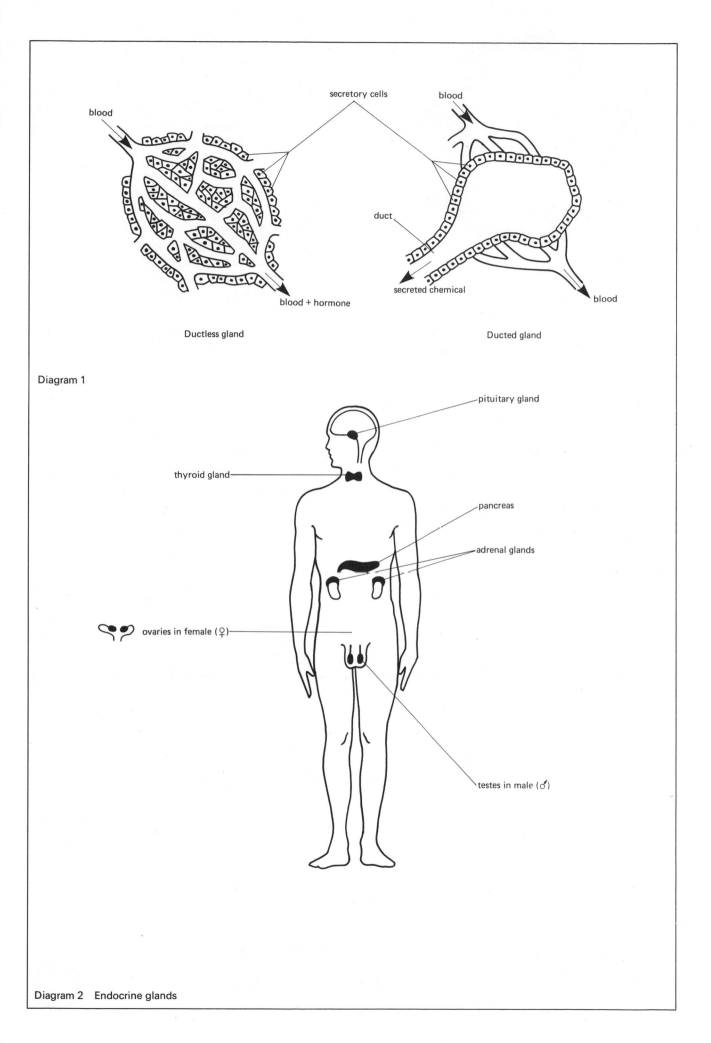

blood

secretory cells

blood

blood

duct

secreted chemical

blood

blood + hormone

Ductless gland

Ducted gland

Diagram 1

pituitary gland

thyroid gland

pancreas

adrenal glands

ovaries in female (♀)

testes in male (♂)

Diagram 2   Endocrine glands

36
Plant responses

Animals can respond to changes in their surroundings by the activity of the endocrine system and the nervous system. Plants also respond to their surroundings but their responses are usually more gradual. For instance, the stem of a plant grows upwards and towards the light; the roots usually grow downwards. The plant is therefore receiving a STIMULUS from the ENVIRONMENT and responding to it. The system which causes this response is different from that in animals and is only recently beginning to be understood.

When a plant grows in a certain direction as a result of a stimulus from its environment, the growth movement is called a TROPISM.

A phototropism is a growth movement caused by the stimulus of light.

A geotropism is a growth movement caused by the stimulus of gravity.

Other tropisms include hydrotropism when the stimulus is water, and chemotropism when the stimulus is a chemical.

When the plant grows towards the stimulus, this is called a positive tropism. If the plant grows away from the stimulus, this is called a negative tropism.

Consider these statements:

1 The shoot of a plant is positively phototropic. 'Positive' means towards, 'photo' means light, so this means that the shoot grows towards the light.

2 The shoot of a plant is negatively geotropic. 'Geo' means gravity and 'negative' means away from, so this means that the shoot grows away from gravity. Since gravity pulls downwards, away from gravity is upwards.

The mechanism of tropisms

Biologists still do not know exactly how all tropisms work. However the series of experiments shown opposite provide some of the evidence for how tropisms work.

In 1880, Charles Darwin, famous for his study of evolution, designed some experiments using grass seedlings. The experiments (see diagram 1) showed that it is the tip of the shoot which is sensitive to the stimulus of light and that the growing region is below the tip.

Thirty years later, experiments by P. Boysen-Jensen (see diagrams 2 and 3) showed that the tip produces a chemical which passes down the plant to the growing region. This chemical causes the cells in the growing region of the plant to elongate (grow lengthwise). The chemical is a plant hormone which is now called auxin.

Sixteen years later, F. W. Went (see diagram 4) collected some auxin by standing the cut-off tip of a seedling on an AGAR block for a few hours. He then placed the auxin-soaked agar on various parts of the shoot. This showed that the side of the plant which receives the most auxin will grow more. This causes the seedling to bend over because one side is growing longer than the other.

Biologists now believe that light shining on one side of a growing plant causes the auxin to gather more on the side furthest from the light. Therefore, more auxin passes down the shaded side of the shoot and causes an increased rate of growth on that side. This increased rate of growth on the shaded side causes the plant to grow towards the light.

Auxins are also involved in geotropisms. If a shoot is lying on the ground, gravity will cause the auxin to collect on the lower side and so the shoot will curve upwards as it grows. However, this does not explain why a root would grow downwards, unless auxin has the opposite effect on root cells, and actually slows down growth. Recent experiments indicate that geotropisms are much more complicated and probably involve several different plant hormones.

Related reading

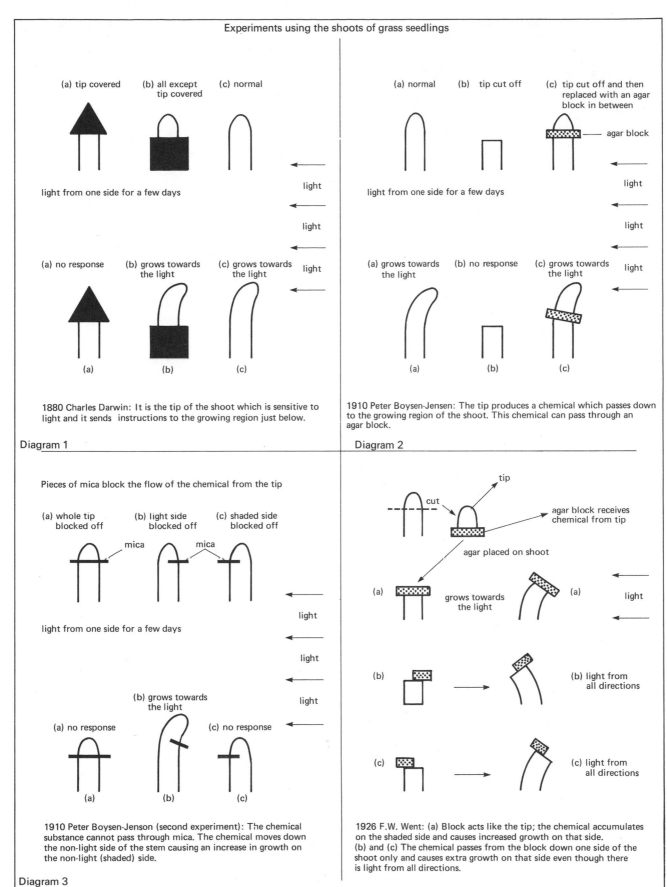

Experiments using the shoots of grass seedlings

Diagram 1

(a) tip covered   (b) all except tip covered   (c) normal

light from one side for a few days

light

(a) no response   (b) grows towards the light   (c) grows towards the light

light

(a)   (b)   (c)

1880 Charles Darwin: It is the tip of the shoot which is sensitive to light and it sends instructions to the growing region just below.

Diagram 2

(a) normal   (b) tip cut off   (c) tip cut off and then replaced with an agar block in between — agar block

light from one side for a few days

light

light

(a) grows towards the light   (b) no response   (c) grows towards the light

light

(a)   (b)   (c)

1910 Peter Boysen-Jensen: The tip produces a chemical which passes down to the growing region of the shoot. This chemical can pass through an agar block.

Diagram 3

Pieces of mica block the flow of the chemical from the tip

(a) whole tip blocked off   (b) light side blocked off   (c) shaded side blocked off

mica   mica

light from one side for a few days

light

light

(b) grows towards the light

(a) no response   (c) no response

(a)   (b)   (c)

1910 Peter Boysen-Jenson (second experiment): The chemical substance cannot pass through mica. The chemical moves down the non-light side of the stem causing an increase in growth on the non-light (shaded) side.

Diagram 4

tip   cut   agar block receives chemical from tip   agar placed on shoot

(a)   grows towards the light   (a)   light

(b)   (b) light from all directions

(c)   (c) light from all directions

1926 F.W. Went: (a) Block acts like the tip; the chemical accumulates on the shaded side and causes increased growth on that side.
(b) and (c) The chemical passes from the block down one side of the shoot only and causes extra growth on that side even though there is light from all directions.

Questions E

The answers to questions 1 to 7 are shown by one of the letters A, B, C, D or E.

1   A nervous impulse passes from one neurone to another across:

A   a synapse.
B   an axon.
C   a medulla oblongata.
D   a myelin sheath.
E   a dendron.

2   When a person touches a hot object, the heat stimulus is detected by:

A   a sensory neurone.
B   a receptor organ.
C   an effector organ.
D   a motor neurone.
E   the spinal cord.

3   The function of the medulla oblongata is to:

A   control precise movement of limbs.
B   store information.
C   promote intelligent behaviour.
D   control breathing movements.
E   increase the rate of metabolism of sugar.

4   The semi-circular canals are organs found in the:

A   eye.
B   ear.
C   nose.
D   intestines.
E   stomach.

5   When the human eye is focused on a distant object:

A   cilliary muscles are contracted.
B   suspensory ligaments are slack.
C   the lens is flatter than normal.
D   the pupil is constricted.
E   the retina moves forward.

6   The hormone secretory organ which has a controlling effect on other hormone glands is the:

A   thyroid.
B   adrenal body.
C   testis.
D   pituitary.
E   Islets of Langerhans.

7   Of the following statements about responses of living organisms the one that applies ONLY to tropisms is:

A   a stimulus triggers the response.
B   the response may be negative.
C   light may be the stimulus.
D   living cells are involved.
E   the response is a growth response.

8   Copy out and then complete the following sentences.
(i)   A mammal can maintain its balance because changes in position are detected in the ear by ....
(ii)   The brain and spinal cord together form the ....
(iii)   Automatic responses to external stimuli, not involving conscious effort, are called ....
(iv)   The part of the brain that co-ordinates muscle action to give balance is called ....
(v)   The substances that are secreted directly into the blood stream and control many body functions are called ....
(vi)   The four taste sensations are sweet, bitter, sour and ....
(vii)   A shoot is said to be positively phototropic because it grows....................................
(viii)   The part of the eye which gives it its colour is ....

9   Explain the difference between the following:
(i)   A nerve and a nerve cell (neurone)
(ii)   The spinal cord and the spinal column
(iii)   Over-activity and under-activity of the thyroid gland
(iv)   The sclerotic and the cornea
(v)   Taste and smell
(vi)   Ducted and ductless glands

10   Write about six lines on FIVE of the following:
(i)   Middle ear
(ii)   Auxins
(iii)   Cerebral hemispheres of the brain
(iv)   Nerve cells

(v)   Sense of taste
(vi)   Accommodation in the eye
(vii)   The pituitary gland
(viii)   Geotropism
(ix)   The nervous system of an invertebrate
(x)   The spinal cord

11   Write an account about twenty-five lines long on ONE of the following. Diagrams may improve the answer.
(i)   The brain
(ii)   Reflex actions
(iii)   Hormones
(iv)   Phototropisms
(v)   Hearing

12

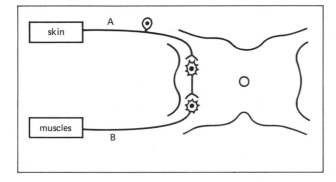

The diagram represents a section through the spinal cord showing a possible pathway for a reflex action.
(i) Copy this diagram and mark on it the following:
   (a) Arrows showing the direction of the nerve impulse.
   (b) A letter S where a synapse would be found.
   (c) A letter W where the white matter would be found.
   (d) A letter G where the grey matter would be found.
(ii) (a) What type of nerve is labelled A?
   (b) What type of nerve is labelled B?
(iii) What structures would you find surrounding the spinal cord?
(iv) Give two examples of a reflex action in humans.

13   (i)   Draw and label a diagram to show the structures of the ear of a mammal.
(ii)   Describe how the structures you have drawn enable you to hear.
(iii)   What part(s) of your diagram are involved in maintaining balance?

(WJEC)

14   (i)   Name the parts A to L indicated in this simple section through the eye.

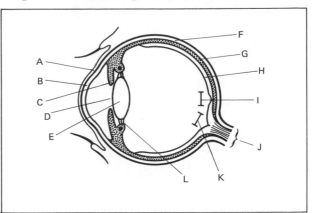

(ii)   Which part of the eye:
   (a) receives the stimulus of light?
   (b) controls the amount of light entering through the pupil?
   (c) is mainly responsible for the refraction (bending) of light entering the eye?
(iii)   When the eye is viewing a near object the ciliary muscles contract. What effect does this have on:
   (a) the suspensory ligament?
   (b) the shape of the lens?
(iv)   The eye is one of the sense organs of man. List the other four sense organs.

15

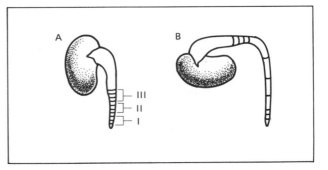

A young bean shoot was marked as shown in diagram A then replanted horizontally and left in a plant pot for two days. It was then removed and observed to be in the condition shown in diagram B.
(i) What has happened after two days to the lines:
   (a) in regions I and III?
   (b) in region II?
(ii) What does this tell you about:
   (a) the region of growth in the root?
   (b) the influence of gravity on the root?
(iii) What is this response called?

37
Movement

All living things move. Animals, like horses, move to eat and to find a mate with whom to reproduce. Plants, like honeysuckle, grow upwards towards the light for PHOTOSYNTHESIS. Flowering plants move their flowers in such a way that they may be POLLINATED by the wind or by insects.

Movement in living ORGANISMS varies from small movements of the inside of cells to the movement of the whole organism from place to place. Movement from place to place is called locomotion and is found in almost all animals. Locomotion is also found in microscopic water plants.

Movement in microscopic plants and animals

These organisms can be seen in any drop of pond-water on a microscope slide. Movement is often quite fast and very varied, but is of two main types:

Movement by cilia and flagella (see diagrams 1 and 2)

Many small organisms move by means of hair-like structures which beat like paddles to move the organism through the water. Some organisms have many small hair-like structures called CILIA and others have fewer, larger structures called FLAGELLA.

Amoeboid movements (see diagram 3)

Amoeba is a jelly-like organism which moves by flowing along slowly. The white blood cells of man can move between the body cells and destroy bacteria by flowing around them. This type of movement is similar to that of the amoeba and is therefore called amoeboid movement.

Movement in large plants

Land plants do not move from place to place. However they must move parts of the plant body to reach light, water and carbon dioxide for photosynthesis. Also these plants reproduce by transferring POLLEN from flower to flower by wind or insects, so the flowers must be moved to a favourable place for such pollination. Plant seeds must be moved away from the parent to a suitable place for GERMINATION.

Plant movements are of two basic kinds: either a slow growth movement, or a movement caused by change of water content of the cells. The movement is the result of a STIMULUS from the ENVIRONMENT. When the stimulus has a definite direction, such as light, then the movement is called a TROPISM (see chapter 36). When the stimulus is general, such as temperature, shock, day-length or light intensity, then the movement is called nastic movement.

Examples of nastic movements

(i) Crocus flowers close at night and open during the day. This is a nastic movement in response to a change in light intensity and a change in temperature between day and night. When the petals are in the light and warmth of the day-time, the cells on the upper surface grow more quickly than the cells on the lower surface, so the petal bends downwards and the flower opens. The opposite movement happens when night brings a drop in light intensity and temperature. This is a nastic movement by growth.

(ii) The mimosa plant shows a nastic movement in response to the stimulus of shock. When the plant is injured or struck, the leaves fold and droop (see diagram 4). This is a nastic movement caused by the rapid loss of water from certain cells in the plant. The mimosa gradually regains its normal erect position.

Movement of large animals

Many-celled animals, like earthworms, insects and man, move by means of muscles contracting. These muscles pull against skin or bone-like supports of the animal body. This is discussed further in chapters 38, 39 and 40.

Related reading

Chapter 36, Tropisms
Chapters 38, 39, 40, Movement in large animals
Chapter 29, Response to stimuli by plants

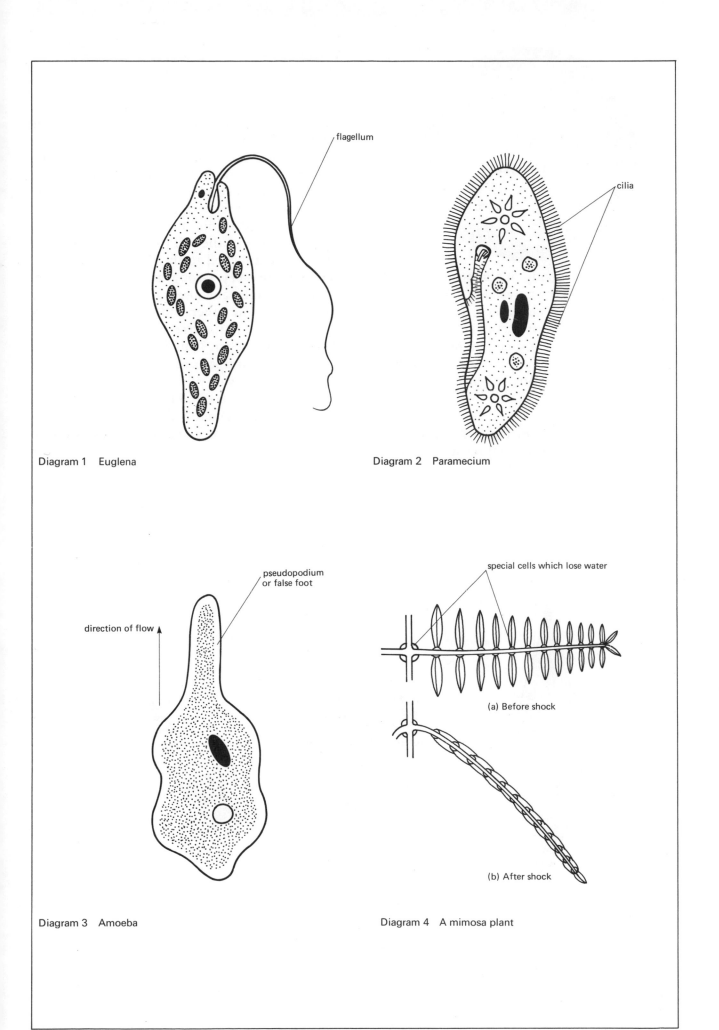

flagellum

Diagram 1   Euglena

cilia

Diagram 2   Paramecium

pseudopodium
or false foot

direction of flow

Diagram 3   Amoeba

special cells which lose water

(a) Before shock

(b) After shock

Diagram 4   A mimosa plant

38
Muscles

The muscles in the body of a large animal act in such a way as to allow movement. The ability to move usually has to be learned, for instance, a young baby has to learn to crawl and then walk. The muscles must act in a certain order to make balance and movement possible and this order is controlled by the nervous system.

Types of muscles

Muscles are the 'meat' of the body and consist largely of protein. There are three main types of muscle: voluntary, involuntary and cardiac muscles.

Voluntary muscles (see diagram 1a)

These muscles, as the name suggests, are under conscious control. The animal can decide when and how much to move them. Voluntary muscles form the flesh of the body and are attached to the skeleton. For this reason they are sometimes called skeletal muscles. Each muscle is a number of bundles of muscle fibres held together in a sheath of connective tissue (see diagram 1b). Each muscle fibre is a muscle cell and looks striped or striated due to the presence of dark and light areas in the cell (see diagram 1c). The muscles are attached to the bones by TENDONS. Tendons are very strong and do not stretch. They are firmly anchored at one end into the bone, and at the other end into the muscle. A muscle can be STIMULATED by a nerve to become shorter and thicker. This is called contraction. Muscles can contract and pull but they cannot expand and push. When a muscle has contracted, it must relax and it then goes back to its original length, when a different muscle contracts. For this reason most muscles work in pairs. When one contracts, the other is relaxed and vice versa. Because the muscles have opposite actions to each other, they are referred

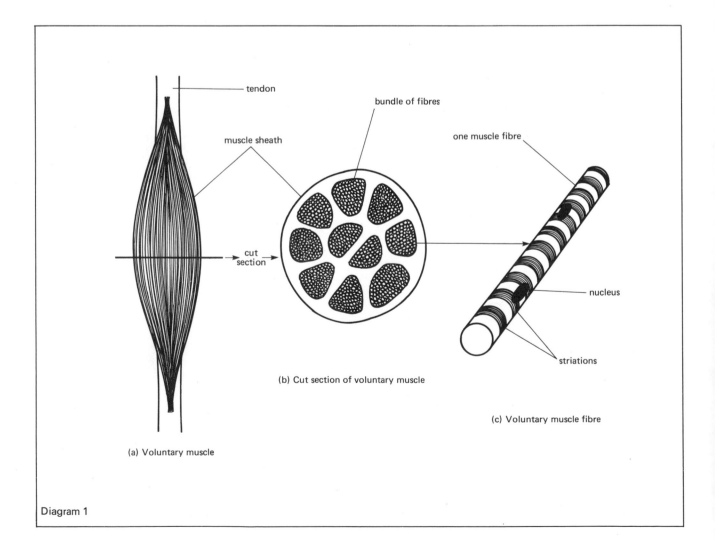

tendon

muscle sheath

bundle of fibres

one muscle fibre

cut section

nucleus

striations

(b) Cut section of voluntary muscle

(c) Voluntary muscle fibre

(a) Voluntary muscle

Diagram 1

to as antagonistic pairs. Diagram 2 shows an example. When the forearm is raised, the biceps muscle contracts and the triceps is relaxed. To lower the forearm, the triceps must contract and the biceps is relaxed.

Antagonistic pairs of muscles control the movement in the leg of an insect in a similar way to that of the human arm, even though an insect has an exoskeleton (skeleton on the outside) rather than an endoskeleton (skeleton on the inside) like man (see diagram 3).

Involuntary muscles

These muscles are not under conscious control and are capable of less powerful actions than the voluntary muscles. They are sometimes called smooth muscles because, when magnified, they do not look striped like the voluntary muscle fibres. Involuntary muscles are in the walls of the gut, where their contractions move the food along by PERISTALSIS. Involuntary muscles also line other hollow organs such as the bladder.

Cardiac muscle

Cardiac muscle is the heart muscle. It looks similar to voluntary muscle but is different in that it is not under conscious control. Cardiac muscle must contract and relax for the whole of the animal's life. It is less powerful than voluntary muscle but it does not become tired.

All muscles must have plenty of food and oxygen carried to them by a good blood supply if they are to work properly. They also perform other jobs in the body apart from locomotion. The diaphragm is a muscle which is essential for breathing. Peristalsis is a muscle action which moves food through the gut. The heart and blood vessels rely on muscle action to keep the blood moving and muscle action is also vital in producing heat which helps control the body temperature.

The control of the muscles by the nervous system is called muscle CO-ORDINATION.

Related reading

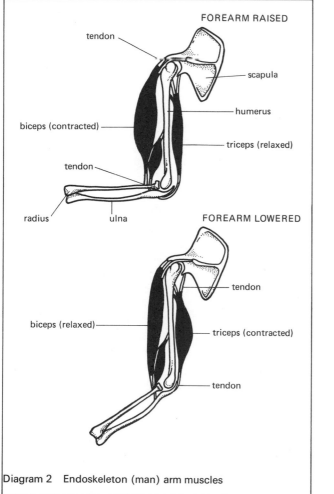

Diagram 2   Endoskeleton (man) arm muscles

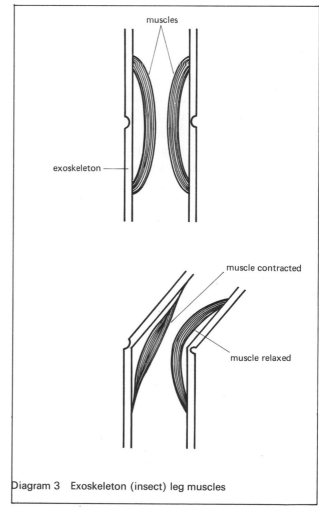

Diagram 3   Exoskeleton (insect) leg muscles

39
The skeleton

The bodies of nearly all animals must be supported or else they would collapse. The animal must also be able to move so that it can capture food or escape from enemies. It is the skeleton of the animal which gives it support, and the joints in the skeleton allow movement.

Animals such as earthworms, caterpillars and slugs are supported mainly by water contained in their cells. This is called a hydrostatic skeleton.

The crustaceans, such as crabs and shrimps, and the insects, such as houseflies and beetles, are supported by a hard skeleton on the outside of their bodies. This skeleton consists of a number of hard plates; movement is possible where the plates join together. This type of skeleton is called an exoskeleton. Larger animals do not have exoskeletons as they would be too heavy and difficult to move. The problems of an exoskeleton would be similar to those found by a man wearing a suit of armour.

All the vertebrate animals have an endoskeleton (see diagram 1). This means that the bones are inside the body. The bones form a supporting framework which gives protection to some of the important organs and offers a firm base on to which muscles can attach.

FUNCTIONS of the human skeleton

Support
The endoskeleton of humans provides the framework on which the body is built. The body would collapse if it were not supported by the skeleton.

Protection
The skeleton protects some of the vital organs of the body. The brain is protected by the skull, and the heart and lungs are protected by the rib cage.

Movement
Many of the bones in the skeleton have ridges or some other structure which offers a firm surface to which muscles can attach. For example, some of the strong back muscles are attached to the processes on the vertebrae and the ridge on the shoulder blade.

Some blood cells are made in the marrow cavities of the larger bones.

The vertebral column (see diagram 2)

The vertebral column, or backbone, consists of thirty-three bones called vertebrae. Each vertebra has the functions of support, protection and movement. The whole backbone is vital to the support of the body and it protects the spinal cord which passes through a hole in the middle of each vertebra. This hole is called the neural canal.

Between each vertebra is a disc of CARTILAGE which allows some movement between the vertebrae, so the whole backbone is flexible. If the backbone were similar to a steel tube it would still give support and protection, but it would be rigid and straight. This would not allow the bending and twisting movements of the backbone of vertebrate animals.

The vertebrae are named according to where they are in the vertebral column. The thoracic vertebrae are in the THORAX and each has a rib attached to it. Lumbar vertebrae (see diagram 3) are the largest of the vertebrae and are found in the lower back region.

Related reading

Chapter 38, Muscles
Chapter 40, Joints

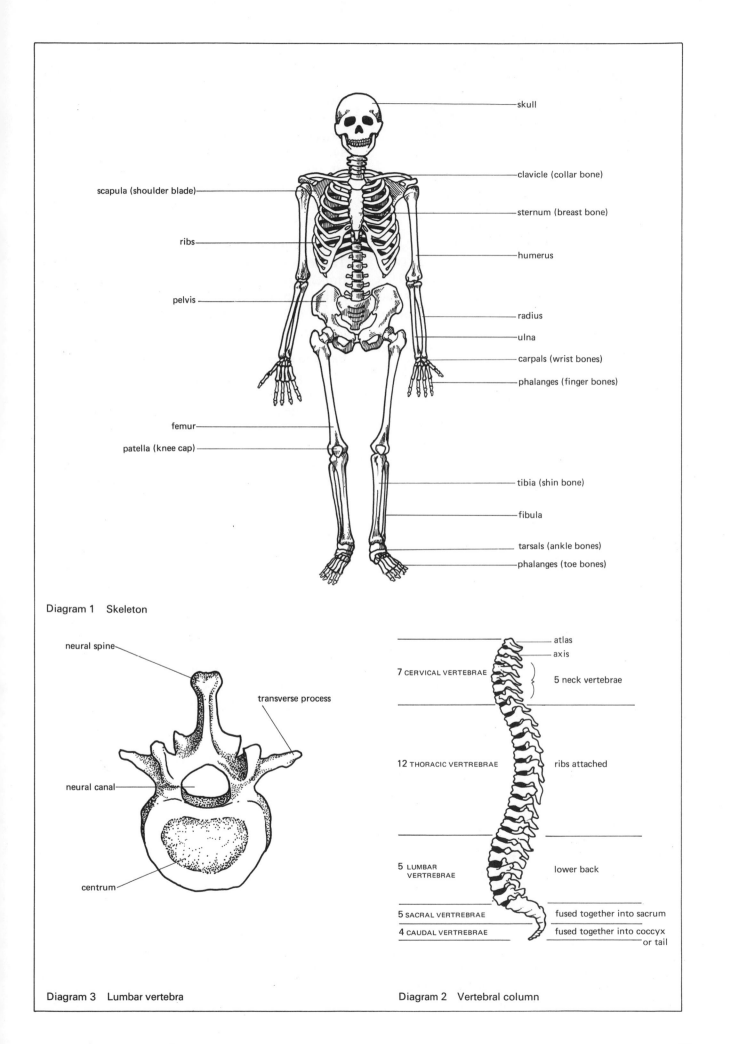

skull

clavicle (collar bone)

scapula (shoulder blade)

sternum (breast bone)

ribs

humerus

pelvis

radius

ulna

carpals (wrist bones)

phalanges (finger bones)

femur

patella (knee cap)

tibia (shin bone)

fibula

tarsals (ankle bones)

phalanges (toe bones)

Diagram 1   Skeleton

neural spine

transverse process

neural canal

centrum

Diagram 3   Lumbar vertebra

atlas

axis

7 CERVICAL VERTREBRAE

5 neck vertebrae

12 THORACIC VERTREBRAE   ribs attached

5 LUMBAR VERTREBRAE   lower back

5 SACRAL VERTREBRAE   fused together into sacrum

4 CAUDAL VERTREBRAE   fused together into coccyx
                             or tail

Diagram 2   Vertebral column

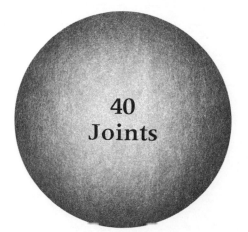

40
Joints

The skeleton of man is made up of about two-hundred bones which are linked together. This framework of bones gives the body support and protection. However, movement would not be possible without joints.

Joints occur within this framework where two bones come together. The joints may allow a little movement or a lot of movement between the bones involved. In the adult, a few joints allow no movement between the bones involved.

Immovable joints (see diagram 1)

Immovable joints occur when two bones are fused together so that no movement is possible. The sacrum is made up of five sacral vertebrae which are fused together into the one bone. Diagram 1 shows a skull which has zig-zag lines, called sutures, across it. When the skull forms in a baby, the various parts are not fused together and this allows the head to be 'moulded' during birth. After a few months, the parts of the skull join together forming these immovable joints called sutures.

Pivot joints (see diagram 1)

The atlas vertebra takes the weight of the skull and the axis vertebra has a small peg which fits into a hole in the atlas. This is called a pivot (or peg-and-socket) joint and it allows the head to be swivelled from side to side.

Gliding or sliding joints (see diagram 2)

These joints occur where two flat surfaces of bone slide over each other. This type of movement occurs between the small bones in the wrist and in the ankle. The movement between each vertebra is also a gliding joint. Each vertebra is separated from the next by a disc of CARTILAGE. These vertebral discs cushion the sliding movement between the vertebrae.

Synovial joints

When there is a lot of movement between two bones, the ends of the bones must be protected or else they would wear away. Joints which allow a lot of movement are called synovial joints. The movement between the bones is lubricated by synovial fluid which is SECRETED by the synovial MEMBRANE. The ends of the bones are covered with a smooth substance called articular cartilage. The joint is enclosed by a capsular LIGAMENT.

Hinge joint (see diagram 3)

The elbows, knees and knuckle joints of the fingers are examples of one type of synovial joint called a hinge joint. A hinge joint can only be raised or lowered, it cannot be moved from side to side.

Ball and socket joint (see diagram 4)

The hip and shoulder are examples of another type of synovial joint called a ball and socket joint. The round head of the femur of the leg fits like a ball into the cup-shaped socket in the pelvic bone of the hip, and the humerus of the arm into the scapula in the shoulder. This type of joint allows the greatest amount of movement of all the joints.

The bones which make the socket must be able to take the force transmitted from the limbs. The pelvis is fused to the base of the spine to form the pelvic girdle and the shoulder blade is bound by muscle to the back of the thorax to form part of the pectoral girdle.

The endoskeleton and the variety of joints of the vertebrate animals give them a wide range of possible movements and great strength. For example frogs, snakes, birds and elephants all have the same basic pattern of skeleton.

Related reading

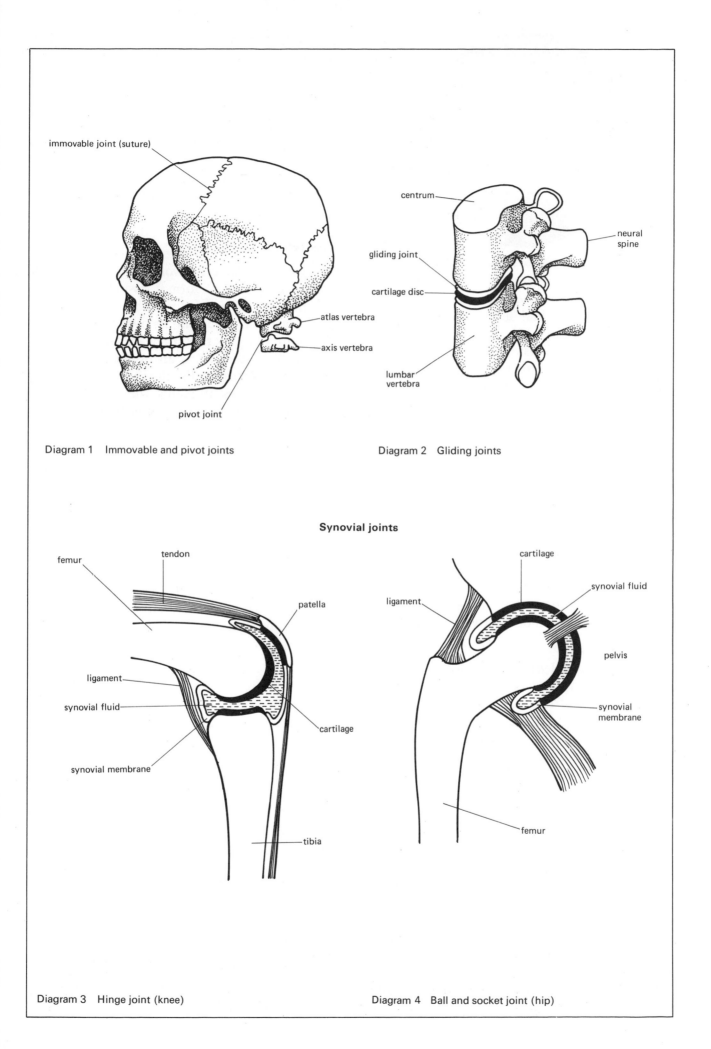

immovable joint (suture)

atlas vertebra

axis vertebra

pivot joint

Diagram 1   Immovable and pivot joints

centrum

neural spine

gliding joint

cartilage disc

lumbar vertebra

Diagram 2   Gliding joints

Synovial joints

femur

tendon

patella

ligament

synovial fluid

synovial membrane

cartilage

tibia

Diagram 3   Hinge joint (knee)

cartilage

synovial fluid

ligament

pelvis

synovial membrane

femur

Diagram 4   Ball and socket joint (hip)

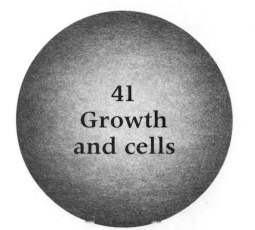

41
Growth
and cells

When living things grow they either increase the number of cells of which they are made, or they increase the size of the cell already present. There is a limit to which any one cell can grow and successfully stay alive. If it is too big it cannot respire, and waste products build up on the inside to a poisonous level. Therefore, most large animals grow by increasing the number of cells in the body.

Increasing the number of cells

Every living cell carries a GENETIC message from the parent which controls the behaviour of that cell. Where there are two parents, both contribute to the message. This message is carried by the cell in the CHROMOSOMES, which are dark threads inside the NUCLEUS. Cell numbers increase by cells dividing in half to make two cells. It is vital that the nucleus should double and then divide equally, so that the genetic message is carried on into each of the two new cells.

Mitotic division

The process of cell division which makes two identical DAUGHTER CELLS is called MITOSIS or mitotic division. The way the chromosomes reproduce and share themselves between the two new nuclei is complicated but can be shown in diagrammatic form (see diagrams). The chromosomes can be seen under a microscope only when a cell is ready to divide and make two daughter cells.

Division of labour

Cells become specialised to do a particular job as animals and plants grow larger. A human must have bone cells, liver cells, muscle cells, nerve cells and cells of many other types. The chromosomes control what type of cell is made and, as growth continues, the cells divide to make more cells of their own type until the body is mature.

The many types of cells make different TISSUE and organs of the body. This specialisation is more efficient than every cell doing every job, and is called 'division of labour'. This is similar to the way people in society specialise in one kind of work, but everyone shares the results of each other's efforts. For example, we do not all grow our own food or smelt our own iron, but we work together to cater for our needs. Muscle cells contract to produce movement, intestine cells absorb food, bone cells are strong for support. They all work together for the needs of the whole body.

Tissues and organs

Groups of similar cells working together at the same job are said to form a tissue e.g. muscles and nerves. When different tissues work together, each contributing its own particular job for a more complex task, they are said to form an ORGAN. The stomach is an organ with muscle cells and SECRETORY cells; the liver is an organ of many tissue types. In plants, the leaf is an organ made up of transporting tissue, photosynthetic tissue and protective tissue.

Injury

An injury will STIMULATE cells to divide and in some animals, such as crayfish and starfish, whole new limbs can be produced this way. In adult humans, however, very little cell division takes place in muscle cells, and none at all has been recorded in nerve cells. A wound is usually healed by scar tissue which is formed by skin cells, and this cannot FUNCTION in the same way as the original tissue.

Related reading

Chapter 42, Patterns of growth
Chapter 53, Genetics
Chapter 4, Cells

Diagrammatic representation of cell division by mitosis in an animal cell with four chromosomes. (Human cells have forty-six chromosomes.)

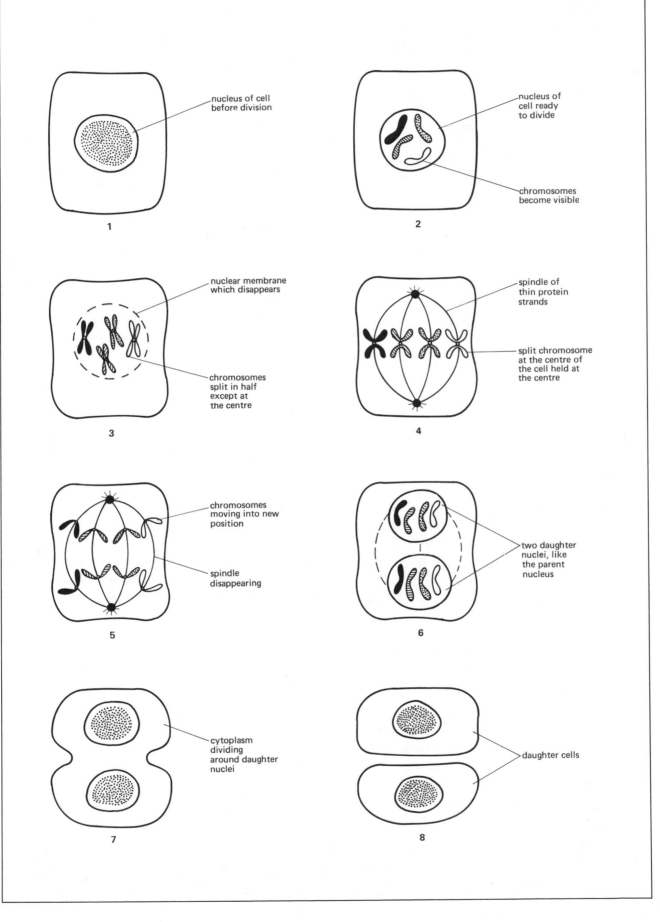

nucleus of cell
before division

1

nucleus of
cell ready
to divide

chromosomes
become visible

2

nuclear membrane
which disappears

chromosomes
split in half
except at
the centre

3

spindle of
thin protein
strands

split chromosome
at the centre of
the cell held at
the centre

4

chromosomes
moving into new
position

spindle
disappearing

5

two daughter
nuclei, like
the parent
nucleus

6

cytoplasm
dividing
around daughter
nuclei

7

daughter cells

8

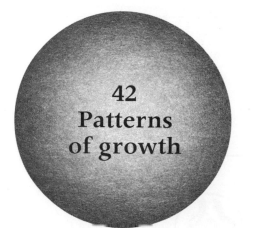

42
Patterns of growth

Growth is a CHARACTERISTIC of all living things and is an increase in the size of an individual. This usually means an increase in the number of cells in the body or an increase in the size of the cells already present. Most ORGANISMS which are made of many cells, such as the mammals and the flowering plants, grow by increasing the number of cells of the body. Growth can also mean the increase in the numbers of individuals in a population.

Conditions needed for growth in individuals

Food
When an animal or plant grows, it is adding new materials to its body. This new material comes from food. A young animal or a young plant will not grow without food.

Removal of poisonous wastes
PHOTOSYNTHESIS and respiration both produce TOXIC or poisonous waste products. These will damage the organism and prevent growth unless they are removed from the surroundings.

Suitable external conditions
Most animals and plants need oxygen and a suitable temperature for the efficient use of food. A man in the arctic may have all the food he needs but if he cannot maintain his body temperature then he will die.

Internal control
Different types of animals and plants grow at different rates and for different lengths of time, even if all other conditions needed for growth are the same. A lobster never stops growing until it dies, whereas a human reaches a certain age at which growth stops. Rats stop growing at about one year old, whereas oak trees continue growing for hundreds of years. The way a living organism grows depends on the GENETIC message, inside the NUCLEUS of each cell, which comes from the parents of the individual. This message is called the GENETIC CODE.

Measurement of growth in individuals

Single individuals of any type are measured for growth either by measurement of a part of the body or by the increase in the size of the whole body. The whole body measurement is usually taken as length or weight. The growth measurements are taken at regular time intervals and are best shown on a graph (see diagram). However, in complex animals and plants the various parts do not always grow at the same rate. A human baby has very short legs and a large head. The legs then grow much faster than does the head until the adult body is formed. Because of this variation in the pattern of growing, weight is usually thought of as a more reliable form of measurement than measuring any particular part of the body.

Measurement of growth in a population

Population growth is of interest to biologists because this growth gives information about the conditions which are controlling reproduction and survival of the young. A population of wild rabbits may be growing very fast; when the reasons for this population explosion are examined, it may be found that the natural enemies of the rabbit, like the stoat, are being killed by farmers.

Population growth is controlled and limited by exactly the same conditions as growth in individuals. These are the availability of food, removal of wastes, external conditions such as weather and internal conditions, such as the numbers of young born to each mother.

If growth were not limited in both individuals and populations, the earth would eventually become overcrowded with giant animals and plants.

Related reading

Chapter 41, Growth and cells
Chapters 55–59, Measurement of population growth

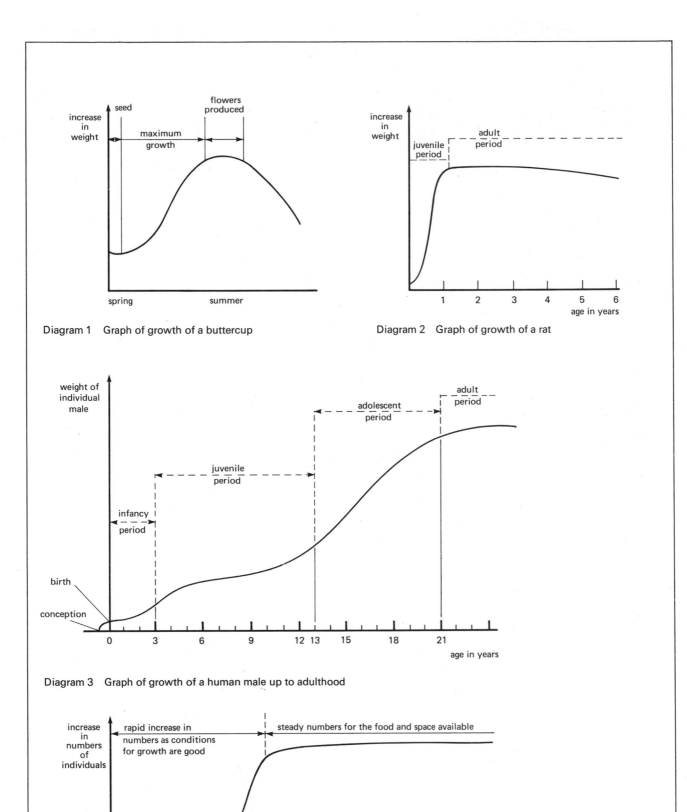

Diagram 1   Graph of growth of a buttercup

Diagram 2   Graph of growth of a rat

Diagram 3   Graph of growth of a human male up to adulthood

Diagram 4   Graph of numbers in a natural population, e.g. sea gulls on an island

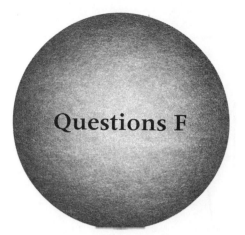

Questions F

The answers to questions 1 to 7 are shown by one of the letters A, B, C, D or E.

1 Of the following, the only non-bony structure is the:

A clavicle.
B choroid.
C ulna.
D patella.
E fibula.

2 Muscles are attached to bones by:

A cartilages.
B tendons.
C muscle fibres.
D ligaments.
E elastic fibres.

3 Of the following, the movement that is NOT caused by muscles is the:

A lowering of the diaphragm.
B passage of mucus up the trachea.
C passage of food through the intestine.
D pumping of the heart.
E swallowing of food.

4 Of the following, the CHIEF characteristic of a ball and socket joint is:

A synovial fluid.
B universal movement.
C articular cartilage.
D capsular ligaments.
E movement in one plane only.

5 Which of the following is NOT a function of the human skeleton?

A Support of the body
B Attachment of muscles
C Protection of internal organs
D Production of blood cells
E Storage of protein

6 Which of the following organisms has a bony endoskeleton?

A An earthworm
B A crab
C An octopus
D A snake
E A leech

7 The process of cell division in skin cells is called:

A meiosis.
B mitosis.
C medulla.
D malleus.
E mitochondria.

8 Copy out and then complete the following sentences:
(i) Euglena swims by using a hair-like structure called a ....
(ii) The biceps and triceps muscles form an .... pair.
(iii) The ribs are attached to the .... vertebrae.
(iv) Muscles are attached to bone by ....
(v) At a joint, bones are attached to bones by ....
(vi) Dark threads in the nucleus of a cell which is about to divide are called ....
(vii) A limiting factor in the growth of a population could be ....

9

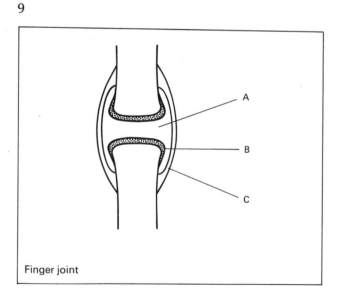

Finger joint

The diagram above shows a finger joint.
(i) Name the fluid present at A.
(ii) State one function of this fluid.
(iii) Name the tissue B which covers the end of the bone.
(iv) State one property of this tissue which is important in carrying out its function in the joint.

104

(v) Name the structure C.

(vi) For movement to occur at this joint two muscles are normally required. What property of muscle makes this essential?

(vii) How does the way in which a human skeleton increases in size differ from the way in which an insect's skeleton increases in size?

(NWREB)

10 Write about six lines on FIVE of the following:
(i)   Involuntary muscles
(ii)   Division of labour of cells
(iii)   Ball and socket joint
(iv)   Locomotion in single-celled animals
(v)   The skull
(vi)   Measurement of growth of an individual
(vii)   The pelvic girdle
(viii) Cartilage
(ix)   Why animals move
(x)   Antagonistic pairs of muscles

11 Write an account about twenty-five lines long on ONE of the following:
(i)   Good conditions for growth in living organisms
(ii) The vertebral column
(iii) Voluntary muscles
(iv) Nastic movements in plants
(v)   Synovial joints

12 (i)   Name three parts of the body which are protected by the skeleton.

(ii) What protects the internal organs of (a) a cockroach (b) a snail and (c) a spider?

(iii) Name three important parts of a synovial joint.

(iv) Give two examples in each case of where the following types of joints occur in the human body:
    (a) Immovable joint
    (b) Gliding joint
    (c) Hinge joint
    (d) Ball and socket joint.

13 The histograms show average human growth, graph A for girls, graph B for boys. Study the graphs and then answer the questions.

(i)   What is the greatest annual increase in height shown by these graphs?

(ii) In what years do girls grow taller more quickly than boys?

(iii) Within the period shown by the graph, at what age would you expect an average girl to achieve her maximum height?

(iv) Other than height, what can be measured to study growth of the body?

(Met. REB)

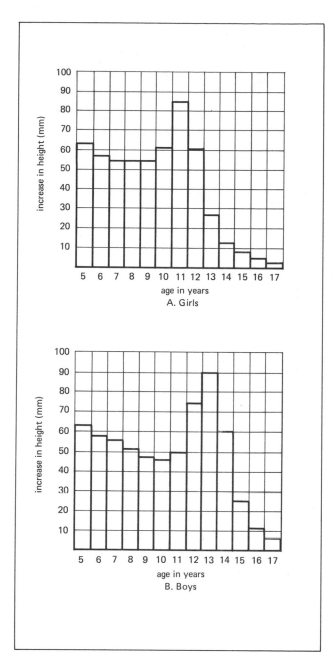

14 (i) Name THREE different types of muscles.

For each type state (a) where it is found, (b) whether or not it is under conscious control and (c) any special properties it may have.

(ii) Carefully describe how an antagonistic pair of muscles is used to raise and lower the forearm (diagrams will improve your answer).

15 Animals are usually capable of greater movement than plants.

(i)   Why is movement from place to place an advantage for an animal?

(ii) Describe briefly how movement takes place in (a) a named animal which has no skeleton and (b) a named animal which has a skeleton.

(iii) Why is movement from place to place unnecessary for a flowering plant?

(WYLREB)

105

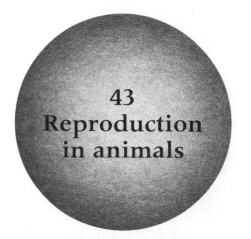

43
Reproduction in animals

All living things must reproduce their kind if they are to survive. Reproduction is the making of a new individual.

Asexual reproduction

Asexual means without male and female parent-cells. There is only one parent and all the new individuals, called offspring, are identical to that one parent. Asexual reproduction is common in very simple animals, but is not common in the more complex or higher animals.

(i) **Binary fission**   Fission means to divide, and binary means into two. Many one-celled animals, for example Amoeba (see diagram 1), divide into two parts and each part grows to the size of the parent.

(ii) **Budding**   Some small many-celled animals, for example Hydra (see diagram 2), reproduce asexually by growing buds. These buds break off at a certain stage and become separate individuals.

(iii) **Parthenogenesis**   The complex animals such as the vertebrates do not normally reproduce asexually. However, some insects such as the greenfly (aphids) can produce offspring from only one parent. This is called parthenogenesis and is the development of young from an egg which has not been fertilised by a male.

Advantages of asexual reproduction

(i)   Every new individual is exactly like the one parent. If the parent is surviving well in a particular place, then the offspring will survive well also.

(ii)   Because there is only one parent, there is no need for a meeting of two adults to produce a new individual.

Sexual reproduction

Sexual means the forming of a new individual which has two parents. Each parent produces a specialised cell, called a sex cell or GAMETE. One male gamete and one female gamete join, or fuse,

to form the first cell of the new individual. This first cell, which is the result of sexual reproduction, is called a ZYGOTE. The joining, or fusing, of the male and female gametes is called FERTILI-SATION. After fertilisation the zygote begins to grow into an embryo.

Many animals produce both male and female gametes in the same adult. These types of animals are called hermaphrodites. However, self-fertilisation is usually prevented by the gametes not being ready to fuse at the same time. In a garden snail, which is hermaphrodite, the sperm (or male gametes) are not produced at the same time as the eggs (or female gametes) and so another snail is needed for fertilisation to take place.

For earthworms to reproduce, any two adult worms come together and fertilise each other (see diagram 3).

In vertebrates and arthropods there is an individual, which makes only sperm, called the male. There is also an individual, which makes only eggs, called the female. The male gamete is the gamete that has to travel to the female egg. All animal sperm can swim in water, but in animals which live on land the sperm must be carried to the female by the male, and introduced into the female body (see diagram 4). Once inside the female, the sperm swim to the egg.

Advantages of sexual reproduction

Every offspring of sexual reproduction is a mixture of the characters of the two parents. This produces a varied population, giving nature a chance to select the offspring most suited to the ENVIRONMENT.

Darwin's theory of natural selection depends on this variation produced by sexual reproduction.

Related reading

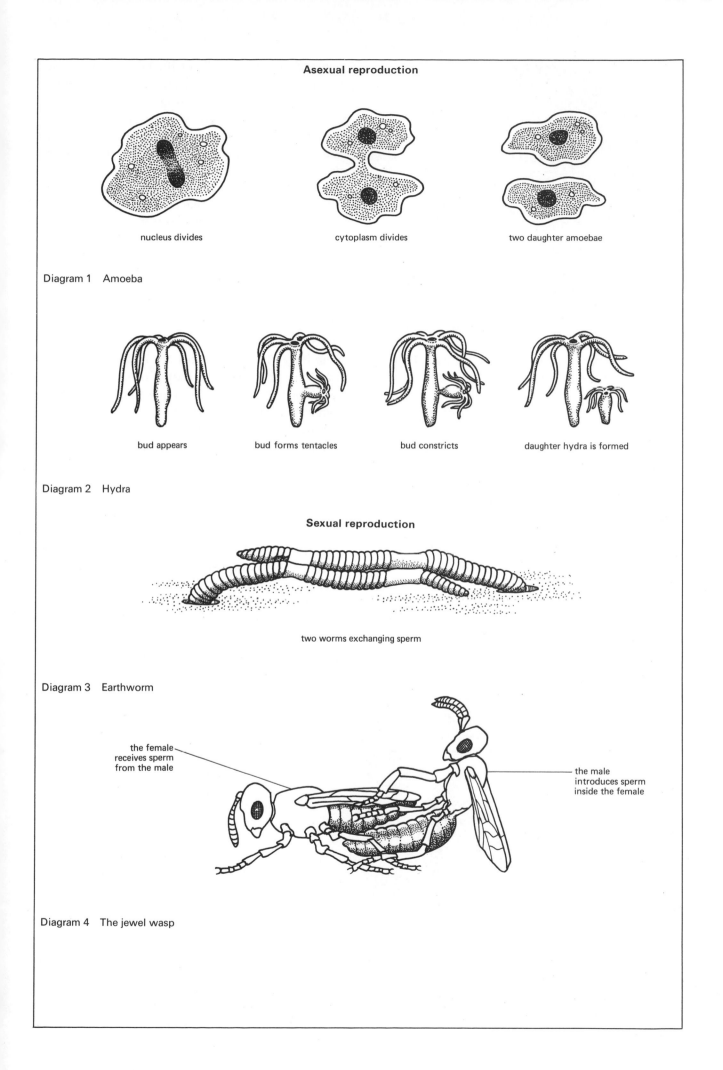

Asexual reproduction

nucleus divides   cytoplasm divides   two daughter amoebae

Diagram 1   Amoeba

bud appears   bud forms tentacles   bud constricts   daughter hydra is formed

Diagram 2   Hydra

Sexual reproduction

two worms exchanging sperm

Diagram 3   Earthworm

the female receives sperm from the male

the male introduces sperm inside the female

Diagram 4   The jewel wasp

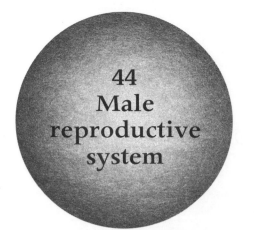

44 Male reproductive system

A male of a SPECIES produces GAMETES, some of which find and then FERTILISE the female gametes. The gametes contain half the CHROMOSOME number of the adult parents so that, when the male gamete fertilises the female gamete, the original chromosome number is restored in the ZYGOTE.

In humans, the sperm contains twenty-three chromosomes, and the ovum contains twenty-three chromosomes. The zygote formed at fertilisation will have forty-six chromosomes, the same number as each cell of the parents. This means that the baby will have a mixture of chromosomes from both of its parents.

In vertebrate animals the sperm are formed in the testes. In water vertebrates, such as fish, the ova and sperm are simply released into the water and the sperm swim to the ova, but this is impossible for land animals. In reptiles, birds and mammals the male has to put his sperm into the female for fertilisation to take place. Unlike a mammal, a baby reptile or bird does not develop inside the female. Instead, the reptile or bird lays the fertilised egg containing a yolk which is a food supply for the growing young.

In mammals, the sperm is transferred to the ovum by passing out of the erect penis of the male, which is placed inside the female vagina. This method of sexual reproduction makes sure that the sperm are so placed that they can swim to the ovum. The fertilised ovum or EMBRYO then grows inside the female uterus until it is ready to be born.

Human male reproductive system (see diagrams 1 to 4)

The testes in man are oval in shape and about 5 cm long. The two testes are in a sac called the scrotum. The scrotum hangs outside the body between the man's legs. This gives the slightly lower temperature which is best for sperm production in the testes.

The testes consist of hundreds of coiled tubules lined with cells which divide to form sperm (see diagram 1). The production of sperm is controlled by HORMONES. The sperm are made continuously from the time a boy is about twelve years old until he is a man of about seventy years. The tubules in the testes join into a coiled tube called the epididymis where the sperm are stored.

When a man becomes sexually excited the erectile TISSUE in the penis becomes filled with blood, causing the penis to become erect and

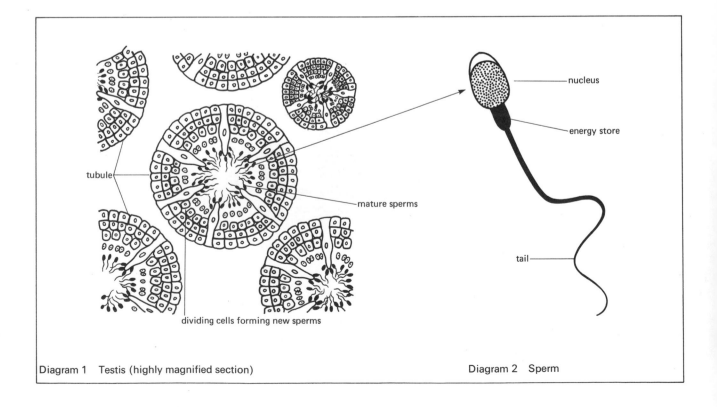

tubule

mature sperms

dividing cells forming new sperms

nucleus

energy store

tail

Diagram 1   Testis (highly magnified section)

Diagram 2   Sperm

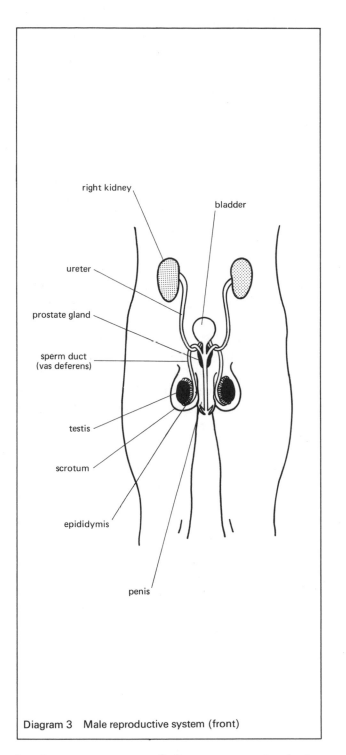

Diagram 3   Male reproductive system (front)

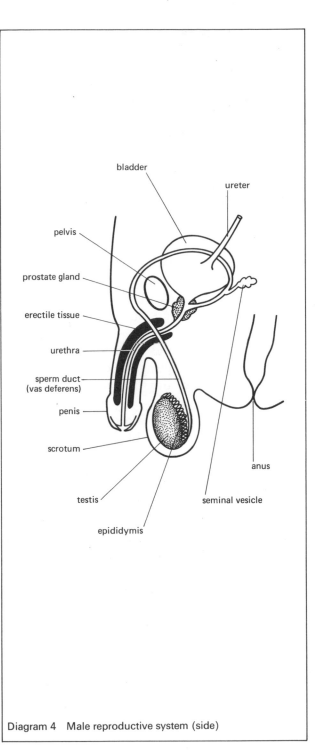

Diagram 4   Male reproductive system (side)

hard. STIMULATION of the erect penis during intercourse or masturbation can cause an ejaculation. An ejaculation is the passing of about 5 cm³ of semen through the urethra. The sperm pass from the epididymis of both testes up the sperm DUCTS. Chemicals SECRETED from the seminal vesicles and prostate gland stimulate the sperm to become active and start swimming. The two sperm ducts join into the urethra so that the semen, which passes out of the erect penis, contains up to 100 million swimming sperm.

The sperm can survive in the female body for two or three days because of chemicals in the semen which feed them. However, only a few

thousand sperm will swim into the oviducts and if an ovum is present only one sperm will fertilise it. As one sperm enters the ovum, an extra MEMBRANE forms around the ovum, stopping the entry of any other sperm.

The NUCLEUS of the sperm fuses with the nucleus of the ovum and a zygote is formed which has a mixture of the chromosomes from each parent.

Related reading

Chapter 45, Female reproductive system
Chapter 46, Pregnancy and birth
Chapter 35, The gonads

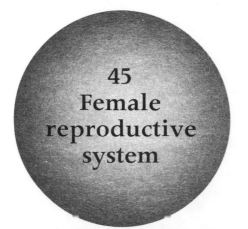

45
Female reproductive system

Many water animals shed their ova into the water. The male sperm released into the water then swim to the ova and FERTILISATION takes place outside the body. This is called external fertilisation. The parents do not usually care for the young.

Reproduction in land animals must involve some method of protecting the ova and sperm so that they do not dry out. This usually involves internal fertilisation.

Internal fertilisation takes place in birds, reptiles and mammals. In birds and reptiles the EMBRYO is then passed out of the body in the form of an egg. The egg contains a yolk which provides food and is surrounded by a shell which protects the young from drying out. Most birds INCUBATE the eggs, keeping them warm in a nest, and feed the young when they hatch by catching food for them. Reptiles do not usually look after their young, although some, for example, alligators, build a nest for the eggs.

After internal fertilisation in mammals, the embryo develops inside the uterus of the female. This means that the embryo is kept in a constant ENVIRONMENT and out of danger. The female can also lead a fairly normal life and does not have to spend a long time sitting on eggs in a nest.

When the young is born, the mother feeds it with milk produced by her mammary glands.

Human female reproductive system (see diagrams 1 to 4)

The two ovaries are oval-shaped and about 3 cm long. Hundreds of thousands of ova are already inside the ovaries of a baby girl when she is born, but the ova are only partly developed.

A girl reaches sexual maturity – or puberty – any time between the ages of ten and fifteen years. At this age, ova in the ovaries begin to develop and each month one ovum is released from the ovary. This is called ovulation. The ovum passes into the funnel of the oviduct which is lined with CILIA. These cilia beat and cause a current which carries the ovum along to the uterus.

The uterus – or womb – is wider and has thicker walls than the oviduct. It is in the uterus that an embryo will develop if the ovum is fertilised. The lower end of the uterus is closed by a ring of muscle called the cervix. The vagina is a tube from the cervix to the outside.

Menstrual cycle

The menstrual cycle is the regular monthly preparation of the uterus to receive a fertilised ovum. An ovum in the ovary develops inside a Graafian follicle which produces a HORMONE. This hormone causes the lining of the uterus to become thicker. The pressure inside the Graafian follicle becomes large so that it bursts, forcing out the ovum. If the ovum is not fertilised it dies. The lining of the uterus then gradually collapses and, about fourteen days after ovulation, the un-fertilised ovum and lining of the uterus pass out of the woman through the vagina. This passing of blood and uterus lining is called menstruation.

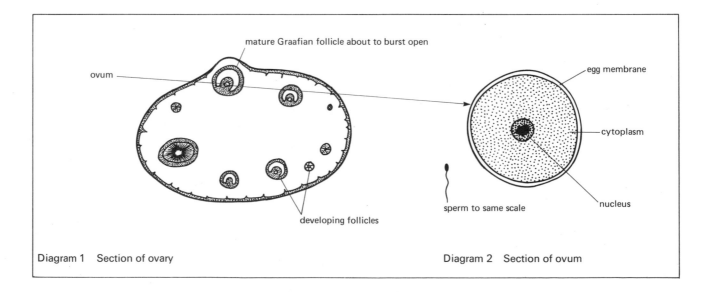

Diagram 1   Section of ovary

Diagram 2   Section of ovum

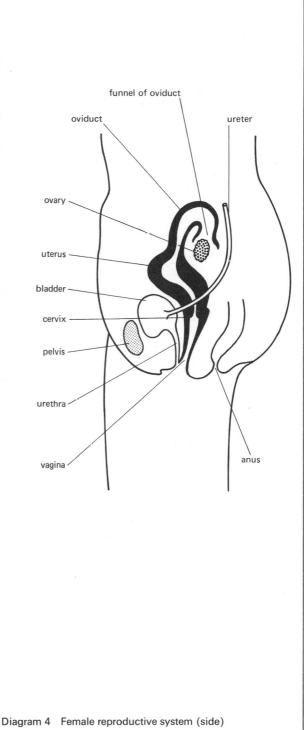

Diagram 3   Female reproductive system (front)

right kidney
oviduct
ureter
uterus
ovary
cervix
vagina
bladder
urethra

Diagram 4   Female reproductive system (side)

funnel of oviduct
oviduct
ureter
ovary
uterus
bladder
cervix
pelvis
urethra
vagina
anus

Women between the ages of about thirteen and fifty years menstruate each month.

Pregnancy can occur if copulation (sexual intercourse) has taken place and there are sperm in the oviduct when the ovum passes through it. One of the sperm may fertilise the ovum forming an embryo.

Hormones SECRETED from the ovary control the lining of the uterus and the development of the embryo. These hormones stop menstruation while a woman is pregnant.

Related reading

Chapter 44, Male reproductive system
Chapter 46, Pregnancy and birth
Chapter 35, The gonads

46 Human pregnancy and birth

FERTILISATION of the ovum by one sperm may take place as the ovum passes down the oviduct. The head of a sperm passes into the CYTOPLASM of the ovum and joins with the NUCLEUS. The fertilised ovum, or ZYGOTE, starts to divide into two, then four, then eight and so on into a ball of several hundred cells (see diagram 1). This EMBRYO of cells moves down the oviduct to the uterus.

By this stage the spongy lining of the uterus has thickened and the tiny embryo sticks to this lining, which then grows around it. This is called the implantation of the embryo. The embryo grows rapidly and a fluid-filled sac develops around it. This is for protection and is called the amniotic sac.

The connection of the embryo with the uterus becomes enlarged to form the placenta, which is made up partly of the mother's TISSUE and partly of the embryo's tissue. Oxygen and food materials pass across the placenta from mother to embryo and carbon dioxide and waste products pass from the embryo to the mother. This takes place by DIFFUSION between the mother's blood, called maternal blood, and the embryo's blood. There is no mixing of the two bloods.

As development continues, the embryo floats freely in the amniotic sac inside the uterus, but it remains connected to the placenta by a special cord called the umbilical cord. After two months, the limbs and the main ORGANS have been

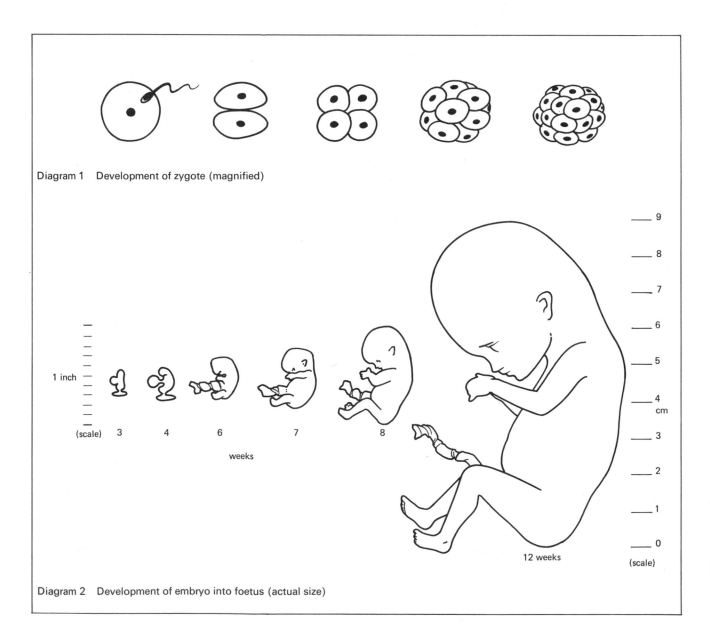

Diagram 1   Development of zygote (magnified)

Diagram 2   Development of embryo into foetus (actual size)

formed and the embryo is now called a foetus (see diagram 2).

During the first three months of pregnancy, the foetus can be damaged by some drugs, by X-rays and by certain diseases such as german measles. Throughout the pregnancy the mother should have a balanced diet including added vitamins and iron and she should not smoke cigarettes.

After three months the foetus is almost completely developed and it then grows steadily without any major changes for the rest of the pregnancy. In humans the time from fertilisation to birth, called the gestation period, is approximately forty weeks or nine months.

Birth

Near the end of the nine months the foetus usually turns head downwards within its amniotic sac (see diagram 3). Birth begins when muscular contractions of the uterus push the head of the feotus down into the mouth of the uterus, called the cervix. The amniotic sac bursts and the flow of fluid can help the feotus to slide out of the uterus and through the vagina (see diagram 4). The muscles of the uterus and vagina stretch to allow the foetus to pass out of the mother. The baby is still attached to the mother by the umbilical cord which is then clamped and cut by the doctor or midwife. The baby takes its first breath and often starts to cry. The rest of the umbilical cord and the placenta, called the afterbirth, come away a little later. The uterus returns to its normal size and shape a few weeks after the birth.

The baby loses a little weight for a few days after birth but should then start to feed on the milk produced in the mother's breasts. The remains of the umbilical cord dry up and fall off leaving a scar, called the navel, on the baby.

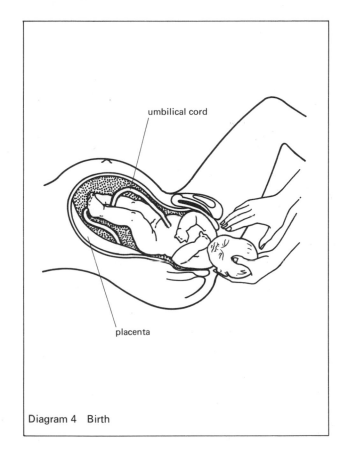

Diagram 4   Birth

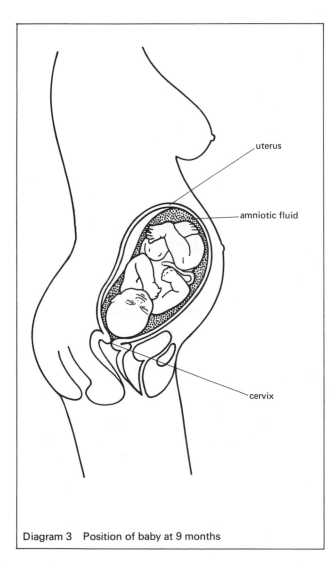

Diagram 3   Position of baby at 9 months

Twins

Identical twins are formed when one ovum, fertilised by one sperm to produce a zygote, divides completely to form two separate groups of cells which produce two identical embryos.

Fraternal, or non-identical, twins develop when two ova are fertilised by two sperm and develop into two embryos.

Related reading

Chapter 44, Male reproductive system
Chapter 45, Female reproductive system
Chapter 9, A balanced diet

47 Metamorphosis

The word metamorphosis means a change of body-form. It is usually used to describe a change which is very easy to see and during which the different body-forms of an ORGANISM live an independent life as if they were different organisms. The best known examples of metamorphosis are found in insects, like the butterfly and in amphibians, such as the common frog. In both these animals the young have a body-form different from the adult. The caterpillar becomes the butterfly and the tadpole becomes the frog.

Metamorphosis in insects

There are two types of metamorphosis in insects, incomplete metamorphosis and complete metamorphosis.

Incomplete metamorphosis

This is a growing process which begins as the young hatch from the egg. These young have no wings and are called nymphs. In some types of insect, like the caddis fly, the nymphs are ADAPTED to live in water and breathe by means of gills. Insects have an exoskeleton and so they grow bigger by stages, shedding the old exoskeleton at the end of each stage. This is called moulting. The nymphs go through this process of moulting many times before they develop wings and become adult.

Locusts develop by incomplete metamorphosis, and the nymphs are known as hoppers. The hoppers moult five times before they develop the wings and sex organs of the adult (see diagram 1). The adult is called an imago.

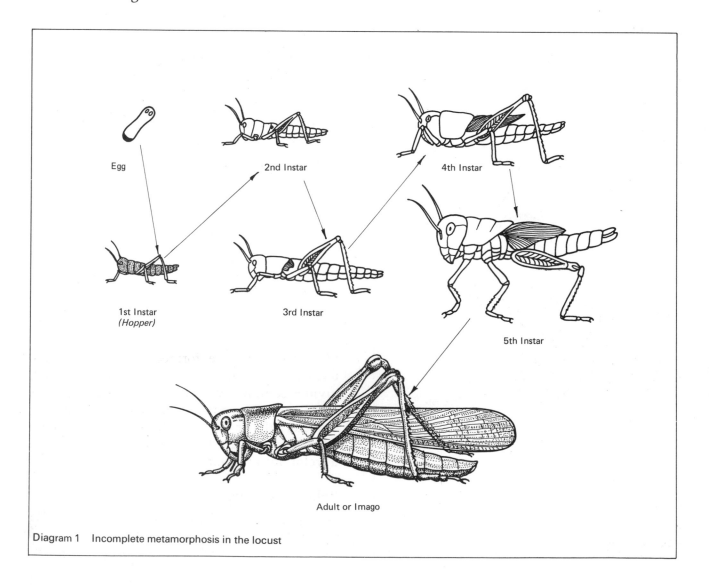

Egg

2nd Instar

4th Instar

1st Instar
(Hopper)

3rd Instar

5th Instar

Adult or Imago

Diagram 1   Incomplete metamorphosis in the locust

114

Complete metamorphosis

In this process the young, when it hatches from the egg, is totally unlike the adult and is called a LARVA. Caterpillars, grubs and maggots are the common names given to some of the insect larvae.

The larva is a simple animal, often without eyes, and devotes its time and energies to eating vast amounts of food and to growing. After a period of growing and moulting, the larva goes through a period of inactivity. It surrounds itself with a cocoon of silk-like material which hardens and protects the insect inside. This is called a pupa and it is inside this pupa that the reorganisation of body TISSUES takes place to produce the complete adult which emerges from the pupa. (See diagram 2.)

Metamorphosis in the amphibians

All amphibians lay their eggs in water and, to help ensure that some are FERTILISED and grow to adults, many eggs are laid at one time. When the young hatch from the eggs they are unlike the adult. The tadpole of the common frog hatches with no mouth and has large gills for breathing. There is a long fish-like tail and no sign of limbs. The change of body-form which results in a tiny version of the adult frog takes place over several weeks, depending on temperature and the food available to the tadpole.

Within three months of hatching from the egg, a young frog has developed limbs, lost its tail and is able to breathe air.

Related reading

Chapter 57, Examples of nymphs and larvae

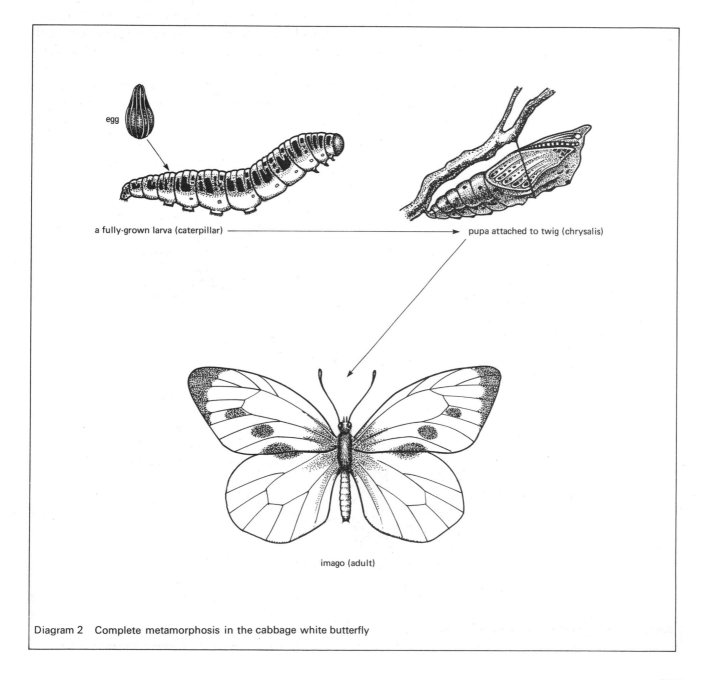

egg

a fully-grown larva (caterpillar) ⟶ pupa attached to twig (chrysalis)

imago (adult)

Diagram 2   Complete metamorphosis in the cabbage white butterfly

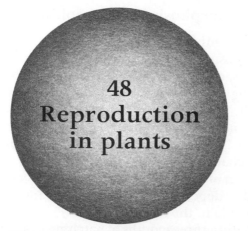

48 Reproduction in plants

Plants, like animals, must reproduce to survive. Many plants can reproduce both sexually and asexually at times in their adult life.

Reproduction in non-flowering plants

The flower is an organ of sexual reproduction but many types of plants do not produce flowers. Examples of such non-flowering plants are mosses and ferns. These reproduce asexually in one generation and then sexually in the next generation. This type of life-cycle is called alternation of generation. (See diagram 1.)

The large green fronds of the fern are the asexual generation of the plant and produce SPORES in brown spots on the back of the frond. These spores fall off the plant and develop in damp soil. The spore produces a tiny green cushion-like plant called a prothallus which is independent of the large spore-producing plant. Prothalli often grow in clusters near the parent plant. The prothallus produces male and female gametes on its underside. The male gametes are released into the soil-water trapped under the plant and swim to a female gamete of a different prothallus. Ferns and other non-flowering plants are dependent on very damp conditions to provide a means by which the male gamete reaches the female gamete.

Non-flowering plants which live in water or in very wet conditions, like seaweeds, reproduce vegetatively by parts of the body breaking off and becoming independent. Sexual reproduction happens when vast numbers of male gametes are shed into the water where just a few will swim to find a female gamete. The female gamete usually remains inside the parent plant until FERTILISATION.

Self-fertilisation is avoided as male and female gametes of the same plant are not ripe at the same time.

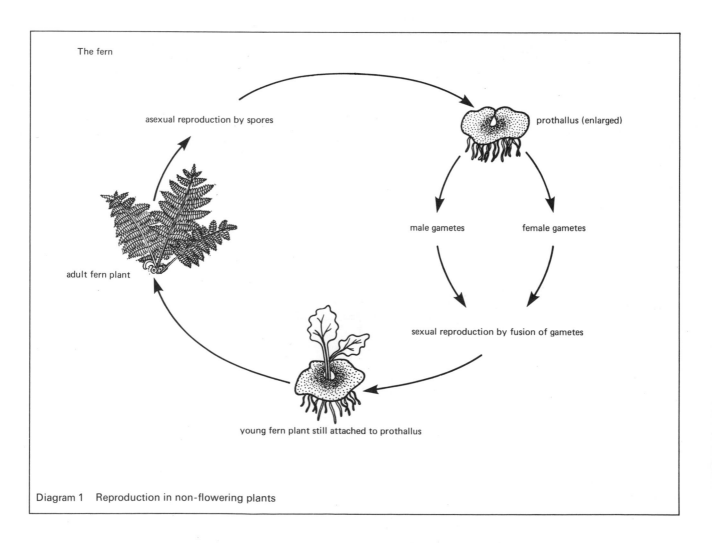

Diagram 1 Reproduction in non-flowering plants

Asexual reproduction in flowering plants

Asexual reproduction is reproduction without the formation of sex cells or GAMETES, and with only one parent. A large mass of cells separates from the parent and becomes independent. This is called vegetative propagation and is used by the plant when conditions for growth are good. Vegetative propagation is often associated with food storage by a plant for survival when the parts above ground die in winter.

Bulbs and corms are a way of survival over winter for plants like the daffodil and crocus, but the formation of these is not always asexual reproduction as a completely new individual is not necessarily formed.

The part of the parent plant which separates and becomes independent may be part of the stem, the root or the leaf, which is then changed or modified.

(i) **Modified stems**   The potato tuber and the strawberry runner are examples of stems which run along the surface of the ground, put down new roots and become new individuals. (See diagram 2a.)

(ii) **Modified roots**   The root system of the dahlia and the lesser celandine form tubers which break off and form new plants. (See diagram 2b.)

(iii) **Modified leaves**   If a leaf of Tolmeia or Begonia becomes detached, small buds form on the leaf, roots grow into the soil and a new plant is formed. (See diagram 2c.)

Sexual reproduction in flowering plants

The flower is the organ of sexual reproduction and this is discussed fully in the next chapter.

Related reading

Chapter 13, Plant storage organs
Chapter 2, The plant kingdom
Chapter 43, Asexual reproduction in animals
Chapter 49, The flower

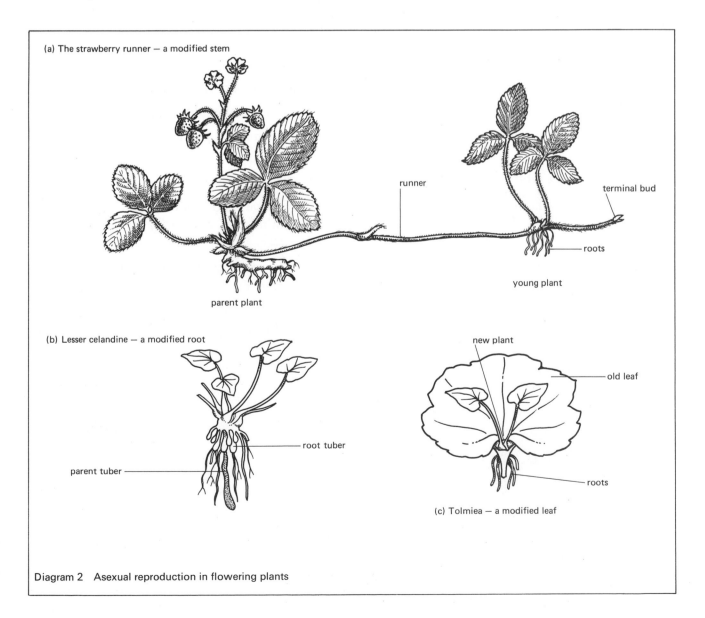

(a) The strawberry runner — a modified stem

runner

terminal bud

roots

young plant

parent plant

(b) Lesser celandine — a modified root

root tuber

parent tuber

new plant

old leaf

roots

(c) Tolmiea — a modified leaf

Diagram 2   Asexual reproduction in flowering plants

49
The flower

Flowers are the sexual reproductive ORGANS of a large group of plants called the angiosperms. When FERTILISED, the flowers of a plant form seeds and develop fruits. These seeds give rise to the next generation of that plant.

There is a great variety of shape, size and colour in different flowers. Most flowers contain a male part, which makes the male GAMETES, called POLLEN. There is also a female part, which makes the female gamete, called the ovule. Fertilisation of the ovule by the male gamete in the pollen forms a seed.

Pollination

Flowering plants do not move around so something else must take the pollen from the anther of one flower to the stigma of another flower. Usually insects or the wind do this, and the action is called CROSS-POLLINATION. Although most flowers cross-pollinate, a few can SELF-POLLINATE. Self-pollination means that the pollen from the anther of a flower fertilises an ovule of the same flower.

An insect-pollinated flower has a different appearance from a wind-pollinated flower although they usually contain the same parts (see table below).

When the pollen grains are on the stigma, pollination is complete. The pollen grains then grow a tube which goes down the style and into an ovule. Fertilisation then takes place (see chapter 50).

Comparison of insect- and wind-pollinated flowers

Insect-pollinated flowers (see diagram 1)

1 Brightly coloured petals which attract insects.

2 Nectar and scent which attract insects.

3 Small amount of large, heavy pollen grains which are often spikey. These stick easily to an insect when it visits the flower.

4 Stamens are within the flower and are firmly fixed so that an insect has to brush against them.

5 Flat, sticky stigma inside the flower so that pollen grains stick to it when an insect, covered in pollen, brushes past it.

Wind-pollinated flowers (see diagram 2)

1 Petals usually small and often green.

2 No nectar or scent.

3 Very large amounts of small, light, smooth pollen grains that can easily be blown by the wind.

4 Stamens PROTRUDE from the flower and are loosely attached so that they shake in the wind.

5 Feathery stigma hanging outside the flower so that it forms a net for pollen grains in the air.

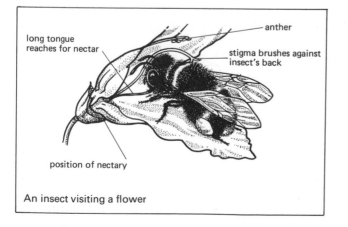

An insect visiting a flower

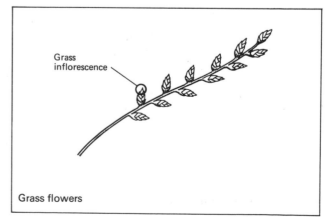

Grass flowers

Parts of a flower (see diagrams 1 and 2)

Receptacle
This is the swollen part of the flower stalk which forms the base of the flower.

Sepals
These protect the flower when it is in the bud. They are usually green.

Petals
The petals of insect-pollinated flowers are usually brightly coloured and attract the insects. They may also have nectaries at their base and produce a scent.

Stamens
The stalk of a stamen is called a filament and the top is called an anther. The pollen grains which contain the male gametes are formed inside the anther. Therefore the stamens are the male parts of the flower.

Carpels
The carpels are the female part of the flower. A carpel consists of three parts:

(i) **Stigma**   This is the top of the carpel to which the pollen grains will stick after pollination.
(ii) **Style**   This is the support for the stigma.
(iii) **Ovary**   This is the swollen base of the carpel in which the ovules develop.

The number and arrangement of these parts depends on the particular type of flower. The sepals or petals may be joined to form a cup or they may even be absent, as in some wind-pollinated flowers. The stamens and carpels may vary in size and number, for example, a buttercup has about sixty stamens and some thirty or forty small carpels. In some flowers, such as dandelions or daisies, what appears to be one flower of many petals is in fact an inflorescence consisting of hundreds of small florets, each floret having petals, stamens and an ovary.

Related reading

Chapter 50, Fruit formation and dispersal
Chapter 51, Seeds and germination

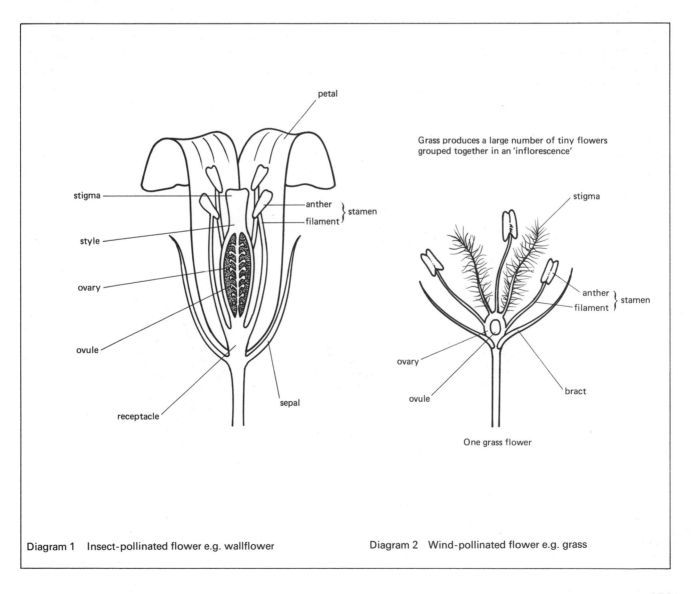

Grass produces a large number of tiny flowers grouped together in an 'inflorescence'

One grass flower

Diagram 1   Insect-pollinated flower e.g. wallflower

Diagram 2   Wind-pollinated flower e.g. grass

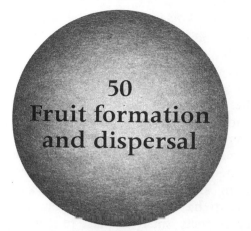

50
Fruit formation and dispersal

POLLINATION is the transfer of the POLLEN grains from the stamens to the stigma.

FERTILISATION is the fusion of the male GAMETE from the pollen grain with the female gamete in the ovule.

After pollination the male gamete in the pollen grain is on the top of the stigma and must travel down to the ovule in the ovary for fertilisation.

Fertilisation

On the top of the stigma are certain chemicals. If any pollen grains of the same SPECIES have landed on the stigma during pollination, the chemicals cause each pollen grain to grow a tube. Pollen from a different species will not develop a tube. The tubes grow down through the style and into the ovary. One of the tubes then grows through a micropyle, which is a hole in an ovule. The male gamete passes from the tube into the ovule and fuses with the female gamete. All of the ovules in the ovary may be fertilised but each needs a separate male gamete in a separate pollen tube.

After fertilisation the petals, stamens, style and stigma of the flower shrivel and usually fall off. Food from the leaves goes to the ovules and ovary. The ovules develop into seeds, and the ovary, or receptacle, develops into FRUITS (see diagram 1). The fruits develop so that the seeds may be removed or dispersed from the parent plant. This prevents overcrowding and competition for light and water.

Dispersal of fruits and seeds

Wind dispersal (see diagram 2a)
The dandelion and clematis fruits have a 'parachute', and the fruits of lime, sycamore and ash have 'wings' which help to carry the seeds away from the parent with the aid of wind.

Self-dispersal (see diagram 2b)
The fruits of the pea family, such as sweet-pea, lupin and gorse, are in the form of a pod. The pod dries in the sun and shrivels and this puts tension on the smooth seeds. The pod then splits open and curls back, throwing out the seeds. A similar mechanism occurs in wallflower fruits, which split open at the base.

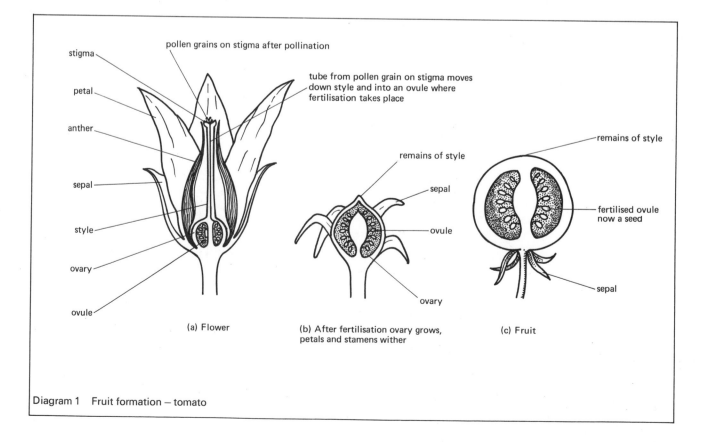

Diagram 1   Fruit formation – tomato

Animal dispersal (see diagram 2c)

Some plants, such as burdock and cleavers, develop fruits which have small hooks on them. These fruits become tangled in the fur of passing animals and so are carried away from the parent plant.

Animal dispersal (see diagram 2d)

Many fruits are juicy and fleshy, which makes them attractive food for birds and mammals. Some species, such as blackberry and strawberry, have small hard-coated seeds which are eaten with the fruit. They pass unchanged through the animal's digestive system and are dispersed along with the FAECES. In other plants, such as apple, plum and cherry, the larger seeds are dropped after the soft, fleshy part of the fruit has been eaten.

Some plants, such as the trees, can survive from year to year and are called perennials. However, many plants grow from a seed, flower, reproduce and die in one year. They are called annuals. Annuals survive because they make seeds which can survive the winter conditions and then grow the following spring. The resting, inactive time of a seed is called dormancy.

Dormancy ends when conditions are right for the seed to grow. The time at the end of dormancy and the beginning of growth is called GERMINATION. Some seeds can remain dormant for many years and then germinate, but most gardeners know that the longer they keep their seeds, the smaller will be the percentage that will germinate.

Related reading

Chapter 49, The flower
Chapter 51, Seeds and germination

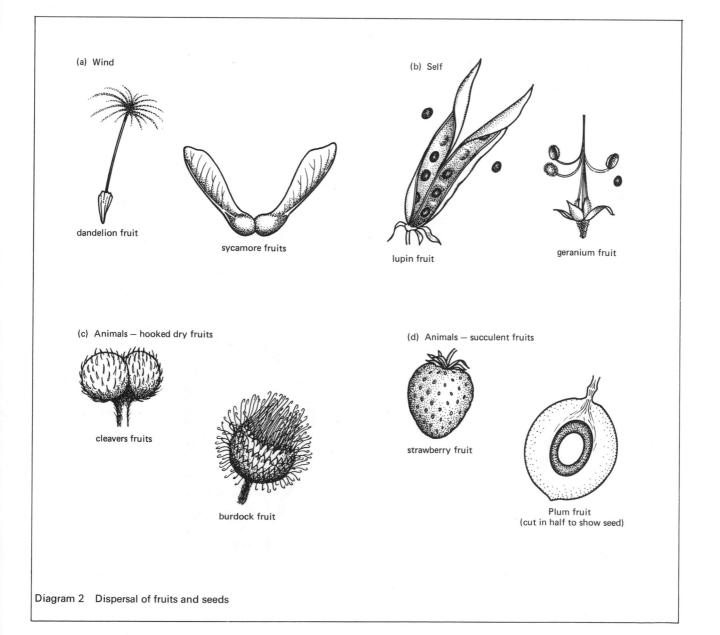

Diagram 2   Dispersal of fruits and seeds

51
Seeds and germination

A seed develops from the FERTILISED ovule of a flower. The seed consists of a plant EMBRYO and a supply of food which will keep the embryo alive until it is able to make its own food by PHOTOSYNTHESIS. The seed is enclosed in a tough seed coat, called a testa, which helps it to survive bad conditions.

While the seed is dormant, it contains very little water. This makes it hard and resistant. The chemical processes of living are slowed down and only small amounts of food and oxygen are used.

Seed structure (see diagrams 1 and 2)

The tiny embryo in the seed consists of a young root, called a radicle, and a young shoot, called a plumule. One or two seed leaves called COTY-LEDONS are also present. In some seeds, such as French bean, the supply of food is stored in the cotyledons but in other seeds, such as maize, the food is in a special store called the endosperm. This food store contains both starch for energy and protein for growth. The embryo and food store are enclosed in the testa.

The hilum is a scar on the testa where the ovule used to be attached to the ovary. The micropyle is the opening through which the pollen tube entered the ovule during fertilisation in the flower and remains as a small hole in the testa through which water can enter the seed.

Germination

Dormancy, the resting stage of the seed, ends when the seed starts to grow. This beginning of growth is called GERMINATION and conditions must be suitable before a seed can germinate. A seed must have oxygen, plenty of water and a suitable temperature. If any one of these three conditions is not fulfilled, the seed will not germinate.

Simple experiments, using CONTROLS, can be set up to prove that oxygen, water and a correct temperature are needed for germination. The oxygen is needed for respiration, and the water and suitable temperature are needed so that ENZYMES in the seed can act on the insoluble food store and make it available to the growing embryo.

French bean (see diagram 3)
The food store of the French bean seed is in the two cotyledons therefore the seed is called a non-endospermic seed. At the start of germination the French bean seed absorbs water and swells. The radicle then bursts through the testa and grows down into the soil. Water and salts are absorbed from the soil and pass to the rest of the seedling. The hypocotyl then starts to grow and pulls the cotyledons out of the testa and up through the soil. The plumule is protected between the two cotyledons. Once above the soil, the leaves of the

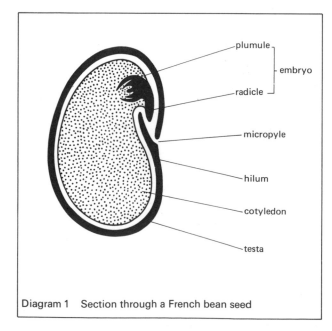

Diagram 1   Section through a French bean seed

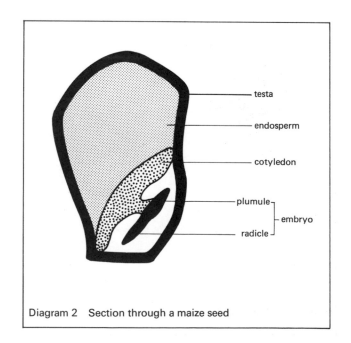

Diagram 2   Section through a maize seed

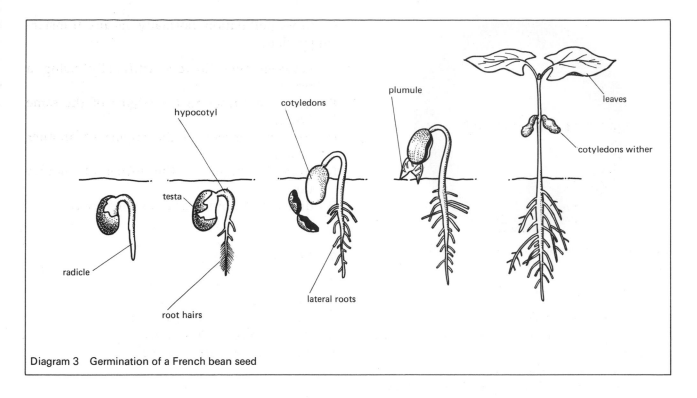

Diagram 3   Germination of a French bean seed

plumule grow and start to photosynthesise. After a few days the cotyledons shrivel and fall off as their food supply of starch and protein has been used up.

Maize (see diagram 4)

The food store of the maize seed is stored in the endosperm. This is called an endospermic seed. At the start of germination the maize seed absorbs water, swells and the radicle grows into the soil. The plumule grows straight up and is protected by the coleoptile which is a sheath with a hard pointed tip. When above ground, the first leaves burst out of the coleoptile. The cotyledon remains below the soil absorbing food from the endosperm which eventually becomes used up and rots.

Related reading

Chapter 49, The flower
Chapter 50, Fruit formation and dispersal
Chapter 68, Control experiments

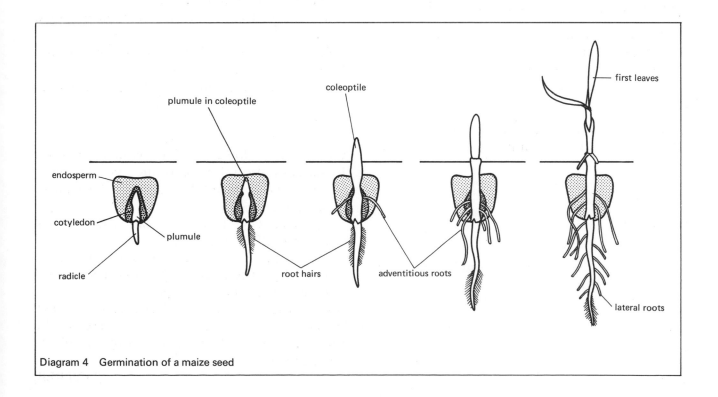

Diagram 4   Germination of a maize seed

Questions G

The answers to questions 1 to 7 are shown by one of the letters A, B, C, D or E.

1 The mammalian placenta is a structure:

A made only of tissues from the embryo.
B where the mother's and foetal blood vessels are close together.
C made only of mother's tissue.
D of muscles to help in the birth of the baby.
E where the mother's blood vessels join the foetal vessels.

2 Wind-pollinated flowers are different from insect-pollinated flowers in that wind-pollinated flowers:

A are brightly coloured.
B have shorter filaments.
C produce larger quantities of pollen.
D produce large quantities of nectar.
E are scented.

3 The natural production of new plants from a root, stem or leaf of a parent plant is called:

A hermaphroditism.
B self-pollination.
C vegetative propagation.
D cross-pollination.
E grafting.

4 Of the following young animals, the one which is most like the adult is the:

A caterpillar.
B maggot.
C chrysalis.
D tadpole.
E calf.

5 Which of the following would a biologist describe as a fruit?

A Potato
B Tomato
C Carrot
D Cabbage
E Onion

6 Cross-pollination normally means transferring pollen:

A between two flowers artificially using a brush.
B from the anther to the stigma of the same flower.
C from the anther to the stigma of another flower.
D from the stigma to the anther of another flower.
E direct to the ovule of the same flower.

7 The epididymis is:

A a hormone-secreting gland.
B the top layer of the skin.
C a coiled tube in the scrotum.
D the entrance to the uterus.
E a tube which carries urine.

8 Copy out and then complete the following sentences:
(i) Sexual reproduction involves the fusion of special cells called male and female.....................
(ii) In a germinating seed, stored starch is changed to a simple sugar by...........................
(iii) A seed is formed from a fertilised ....
(iv) The embryonic shoot of a seed is called
....
(v) An animal having both male and female reproductive organs is called ....
(vi) The first cell of an individual formed by sexual reproduction is called ....
(vii) The seed leaves which store food are called
....
(viii) The resting stage between caterpillar and butterfly is called...
(ix) Asexual reproduction in ferns produces
....

9 (i) Many flowering plants which reproduce by seeds are also able to reproduce 'vegetatively'.
  (a) What is meant by the term 'vegetative reproduction'?
  (b) Name ONE flowering plant which reproduces itself vegetatively.
  (c) Describe, with the aid of labelled diagrams, how the plant which you have named reproduces itself in this way.
(ii) Describe, with the aid of a diagram, ONE method used by gardeners to increase the number of a NAMED plant by division of the parent plant.

(WYLREB)

124

10 Write about six lines on FIVE of the following:
(i)   Gametes
(ii)   Zygote
(iii) Sperm production
(iv)   Puberty
(v)   Alternation of generation
(vi)   Larvae
(vii) Stamens
(viii) Carpels
(ix)   Seed dispersal by animals
(x)   Cotyledons

11 Write an account about twenty-five lines long on ONE of the following:
(i)   Asexual reproduction
(ii) Fertilisation
(iii) Menstrual cycle
(iv) Metamorphosis
(v)   Germination

12 The diagram below shows the life-cycle of a flowering plant. The various stages are marked in sequence as 1, 2, 3, 4, etc.

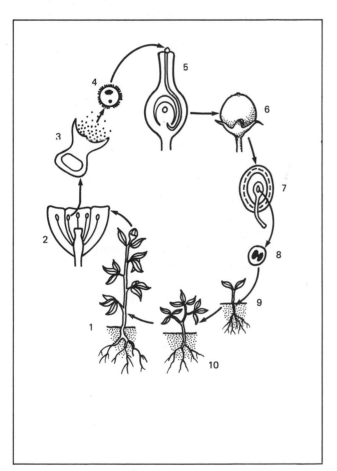

(i)   Suggest names for each of the ten stages, shown in the diagram above.
(ii) Describe what is happening between stages 4, 5, and 6.
(iii) Describe what is happening between stages 8, 9, and 10.

13 (i)   Draw and label a diagram of the male reproductive system.
(ii)   State an advantage that sexual reproduction has over asexual reproduction.
(iii) Why are so many sperm cells released by males during mating?
(iv)   Why are the eggs of mammals smaller than the eggs of birds?
(v)   What is the function of the sac surrounding an unborn baby?
(vi)   From where does an unborn baby receive its oxygen?
(vii) Name one hormone connected with the menstrual cycle.

14 (i)   Describe four differences between insect-pollinated plants and wind-pollinated plants.
(ii) How do insects benefit by visiting flowers?
(iii) Draw a large, fully labelled diagram of a section through a named seed.

(WJEC)

15 The diagram shows the human foetus in the uterus. Write the names of the parts indicated by the letters A to H.

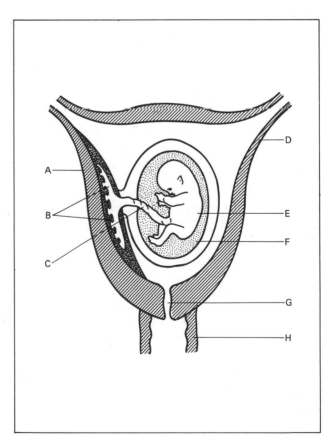

(i)   How does the foetus obtain its supply of food and oxygen?
(ii) How does the foetus dispose of its excretory products?
(iii) Explain how identical and non-identical twins are formed.

52
Heredity

Cats are always expected to give birth to kittens, acorns to grow into oak trees, and humans to produce babies. Animal and plant breeders have used their observations of the way offspring are like their parents to produce suitable flowers, vegetables, farm animals and animals for sport and pets.

However, animal and plant breeders also noted that the young of any two parents were not all alike. It takes many processes of trial-and-error to breed a champion race-horse, even beginning with two champion parents. Sexual reproduction produces variation in the offspring. Some brown-eyed parents can produce some brown-eyed and some blue-eyed children.

The study of the way CHARACTERISTICS are handed down from parents to offspring is called heredity. Heredity was first studied scientifically by a monk called Gregor Mendel who lived from 1822–1884 in mid-Europe. He studied the garden peas growing in a monastery garden.

The work of Gregor Mendel (see diagrams)

It had long been known that sexual reproduction in all living things involved the joining of two GAMETES. These special cells are produced by the male parent and the female parent. When the gametes join they form the first cell of a new individual which inherits the characteristics of the two parents.

Mendel noted that variation in garden peas had given rise to tall pea plants about 2 metres high and dwarf plants only 40 cm high. These two varieties always produced offspring identical to the parents because pea flowers normally SELF-POLLINATE.

Mendel selected a tall plant as one parent and a dwarf plant as the other. He carefully CROSS-POLLINATED the flowers, using a fine paint brush. When these parents produced seeds, he collected them and raised a first generation of offspring.

He called these plants the F_1 generation. He discovered that all the F_1 plants were tall.

The F_1 plants were allowed to self-pollinate and Mendel again collected the seeds and raised a second generation of offspring, called the F_2 generation. This F_2 generation was a mixture of tall and dwarf plants. Mendel counted the F_2 plants and found that there were approximately three times as many tall plants as dwarf plants. This is called a 3 : 1 ratio. He repeated the whole experiment many times with large numbers of plants but always found the same 3 : 1 ratio in the F_2 generation.

Mendel's conclusions from his work

1   Gametes carry a factor for stem height from parents to offspring.

2   When a tall stem factor from one parent is joined with a dwarf stem factor from the other parent, the stem grows tall. This must mean that tallness is dominant over dwarfness. Dwarfness did not show up in the F_1 generation and so is called a recessive factor. Tallness and dwarfness, or any other pair of factors for a particular characteristic, are now called a pair of alleles.

3   Even though a recessive allele may not affect the appearance of an offspring, it is still present in the cells and can be passed to the next generation. Dwarfness was passed from the F_1 generation to the F_2 generation where some of the offspring were dwarf pea plants.

Mendel repeated his work with peas but using flower colour, the shape of peas and the colour of the pea pod. He found the same mathematical results each time.

Many biologists have since expanded Mendel's work and have found it a sound basis for a scientific study of heredity.

Related reading

Diagrams to show Mendel's work

Let T stand for the factor for tallness
and t stand for the factor for dwarfness

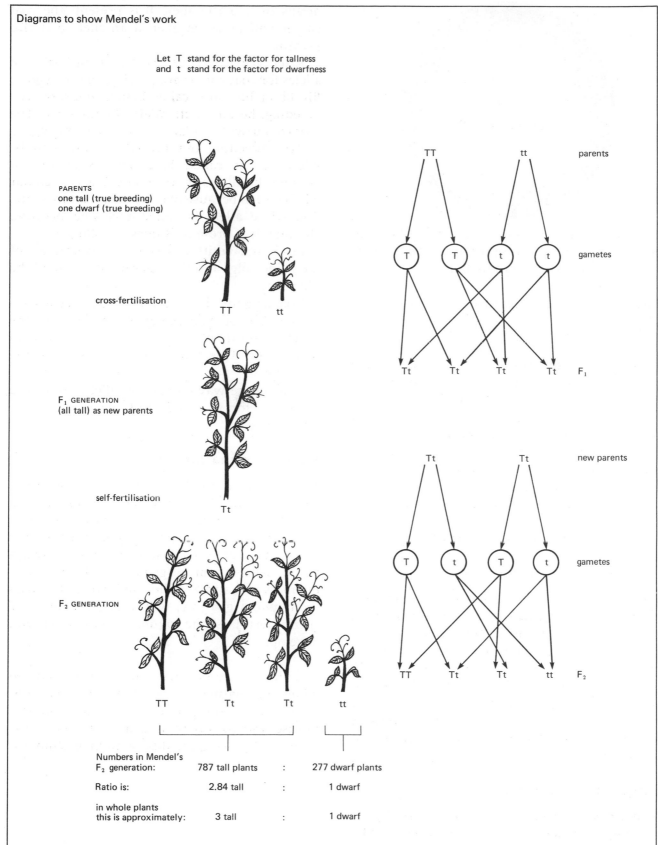

PARENTS
one tall (true breeding)
one dwarf (true breeding)

cross-fertilisation

TT   tt

F₁ GENERATION
(all tall) as new parents

self-fertilisation

Tt

F₂ GENERATION

TT   Tt   Tt   tt

TT   tt   parents

T   T   t   t   gametes

Tt   Tt   Tt   Tt   F₁

Tt   Tt   new parents

T   t   T   t   gametes

TT   Tt   Tt   tt   F₂

Numbers in Mendel's F₂ generation:	787 tall plants	:	277 dwarf plants
Ratio is:	2.84 tall	:	1 dwarf
in whole plants this is approximately:	3 tall	:	1 dwarf

Mendel concluded: where T was present it showed up, so T was dominant
and t was recessive

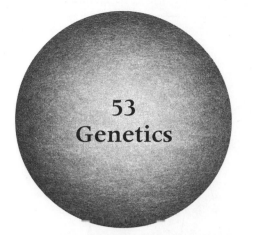

53
Genetics

GENETICS is the scientific study of the way offspring inherit the CHARACTERISTICS of parents as a result of sexual reproduction.

The material of inheritance

It has been shown that the material which controls inheritance is the chemical called DNA. This chemical makes up the dark thread-like structures called CHROMOSOMES. Chromosomes are present inside the NUCLEUS of the cell in pairs, one of each pair coming from each parent. Human cells contain forty-six chromosomes (twenty-three pairs) in every body cell.

DNA contained in chromosomes is transferred from the parents to the offspring inside the GAMETES. These chromosomes then control the appearance and basic behaviour of the individual that grows from the ZYGOTE.

Formation of gametes

When a gamete is formed, only one chromosome from each parental pair of chromosomes passes into the nucleus of the gamete. The nucleus of a gamete always contains half the number of chromosomes that is contained in the adult body cells. This process is called reduction division or MEIOSIS. When a male gamete and a female gamete fuse, the number of chromosomes returns to the full number for the body cells of that ORGANISM.

Single factor inheritance

A chromosome is made up of many GENES. A gene or group of genes controls an inherited characteristic. The gene or group of genes is called the factor or the allele for that characteristic, and is represented in genetics by letters.

Example (see diagrams)
The allele for black coats in guinea pigs is shown by the letter big B. Black coat is dominant to white coat, therefore the allele for white coats is not represented as W but as little b. Dominant

means that, if the allele B is present, the coat colour will be black even if an allele b is also present.

After FERTILISATION the following pairs of alleles for coat colour are possible in the zygote: BB, Bb or bb. BB is called homozygous, or true breeding, because both alleles are the same. The coat colour will be black. bb is also homozygous or true breeding but the coat colour will be white. Bb is called heterozygous or hybrid because the alleles are different. The coat colour will be black because black is dominant to white. The white allele is called recessive because, although it is present, it does not show.

Black and white describe the appearance of the guinea pigs and this appearance is called the phenotype.

Homozygous dominant BB, homozygous recessive bb, or heterozygous Bb describe the genetic make up of the guinea pig with respect to coat colour. This genetic make up is called the genotype.

Tracing the way one pair of alleles controls a certain factor is called studying single factor inheritance.

Sex determination

The sex of an individual is determined by a particular pair of sex chromosomes. In humans, women have a pair of sex chromosomes called XX, and men have a pair of sex chromosomes called XY. Therefore female gametes must always be X, whereas male gametes will be equal numbers of X and Y. When an X sperm fertilises the ovum, a girl (XX) will develop. When a Y sperm fertilises the ovum, a boy (XY) will develop.

Genetics is a science that can help man to breed better animals and plants for food. It can also be used to study the pattern of the inheritance of disease and to advise human sufferers of the chances of passing the disease to their children.

Related reading

Tracing single factor inheritance of coat colour in guinea pigs

Let the allele for coat colour be B for black and b for white.

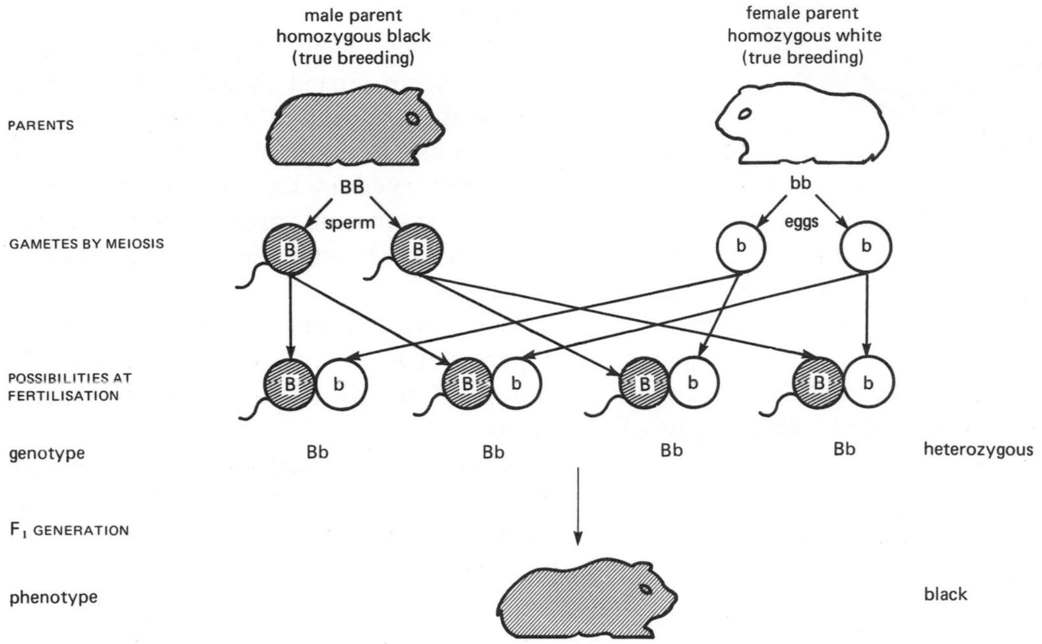

male parent
homozygous black
(true breeding)

female parent
homozygous white
(true breeding)

PARENTS

BB

bb

GAMETES BY MEIOSIS

B   sperm   B   b   eggs   b

POSSIBILITIES AT
FERTILISATION

B b   B b   B b   B b

genotype   Bb   Bb   Bb   Bb   heterozygous

F_1 GENERATION

phenotype   black

All F_1 are the same and are black

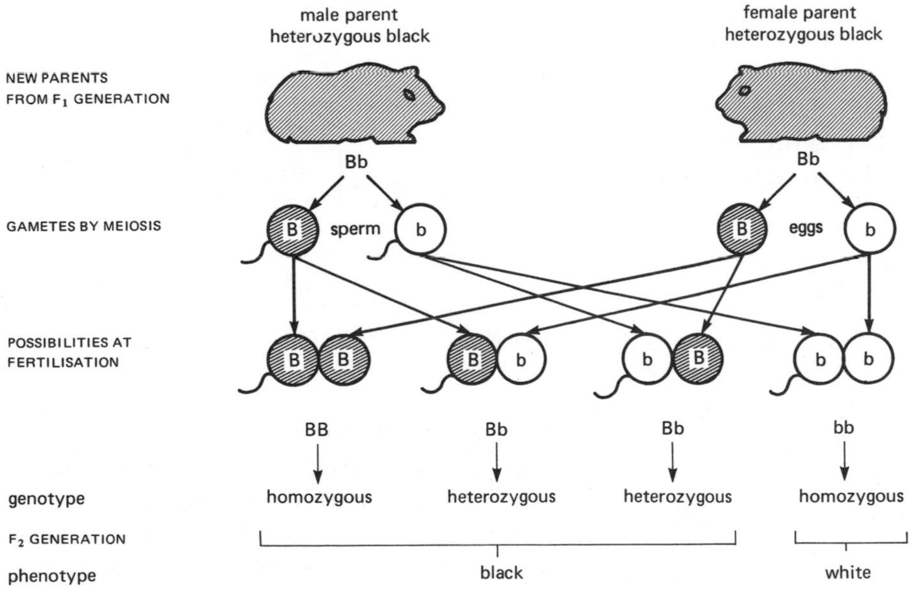

male parent
heterozygous black

female parent
heterozygous black

NEW PARENTS
FROM F_1 GENERATION

Bb

Bb

GAMETES BY MEIOSIS

B   sperm   b   B   eggs   b

POSSIBILITIES AT
FERTILISATION

B B   B b   b B   b b

BB   Bb   Bb   bb

genotype   homozygous   heterozygous   heterozygous   homozygous

F_2 GENERATION

phenotype   black   white

Ratio of F_2 generation   3 black : 1 white

129

54
Evolution

There is a large number of different SPECIES of animals and plants on the earth. It was once believed that these numerous species were all created exactly as they are at the present time.

Charles Darwin (1809–1882) was one of many scientists who disagreed with this and he collected evidence to support a theory of evolution based on natural selection. He published his evidence in a book called 'The Origin of Species'.

Natural selection – variation

In nature, many more young are produced than can be supported by their ENVIRONMENT. For example, one pair of frogs produces hundreds of tadpoles each year, and one oak tree produces hundreds of acorns each year, yet the numbers of frogs and oak trees remain fairly constant.

The young of each species are all slightly different and this variation allows the forces of nature to select only the individuals that are best ADAPTED to their environment. They survive and produce young. Similarly man selects only the best of his farm animals to produce young. In this way gradual changes in the CHARACTERISTICS of a species appear.

Natural selection – mutation

Occasionally an individual which has a marked difference in body form appears in a species. This is called a mutation. Most mutations are not advantageous, but occasionally one occurs which may be of advantage to the individual. The offspring from this individual will inherit this beneficial mutation. In this way a new species may be formed. There has been life on earth for at least five hundred million years, which would be enough time to produce the great number of different species of living things we see about us today.

Modern scientific knowledge of GENES and CHROMOSOMES has supported Darwin's idea of natural selection producing evolutionary change by variation and mutation.

Evidence for evolution

The main evidence that evolution has occurred comes from fossil records, geographical distribution and the comparison of present-day living things.

Fossil records
Fossils were formed when plant and animal remains were buried by SEDIMENTS in lakes and seas, and these gradually changed into sedimentary rocks. Occasionally whole animals have been preserved in ice, such as a mammoth in Siberia, or insects preserved in resin. Also, imprints have been made in rocks by animals and plants.

Many present-day animals and plants do not appear as fossils, and a great number of animal and plant fossils do not look like present-day forms.

Fossil evidence supports the theory of evolution. A geological time scale can be made of the most abundant fossils at different periods of time (see diagram 1).

Geological era	Number of years ago	Dominant animal group
Cenozoic	100 000 000	age of mammals
Mesozoic	200 000 000	age of reptiles
Palaeozoic	300 000 000	age of amphibians
	400 000 000	
		age of fishes
	500 000 000	
	600 000 000	age of invertebrates
Pre-Cambrian		

Diagram 1   Geological time scale

Geographical distribution
Geographical barriers, such as mountains or sea, can lead to some species being found only in certain isolated areas. For example, the marsupials are found only in Australia. They are animals with a pouch, such as the kangaroo. Where marsupials and placental mammals are in competition, the marsupials rarely survive. Geological evidence suggests that Australia separated from Asia before placental mammals evolved and so the Australian marsupials survived as they had no competition.

Comparison of present-day living things

The structure of the brain in all vertebrates is so similar that it suggests that vertebrates have evolved from a common origin. Also the limbs of amphibians, reptiles, birds and mammals are based on the same pentadactyl pattern, even though the limbs have different functions of swimming, flying or walking (see diagram 2).

Related reading

Chapter 52, Variation
Chapter 69, History of biology

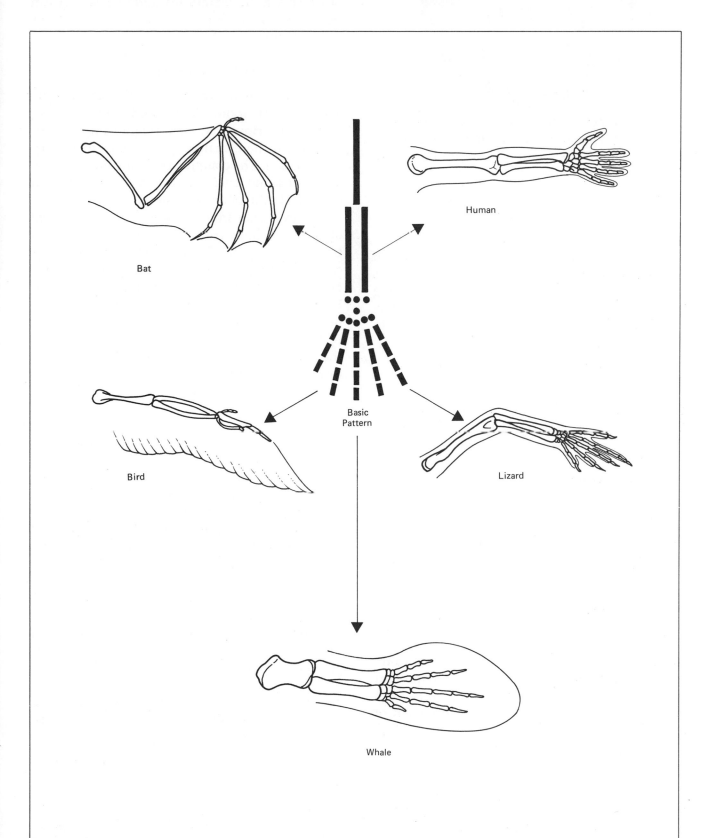

Diagram 2   Pentadactyl limb

55
Ecology

Ecology comes from a Greek word meaning 'home' and it means studying animals and plants in their home. Thus a definition for ecology is the study of ORGANISMS in relation to their ENVIRONMENT.

Environment

An environment means physical conditions such as light and temperature, chemical conditions such as water, oxygen and minerals, and living or biotic conditions which are the inter-relationships of all the animals and plants in the environment.

Habitat

An environment with one particular set of conditions is called a HABITAT. A seashore is a habitat, a freshwater pond is a different habitat.

Adaptations to environment

The animals and plants found in the seashore habitat will not be the same as the animals and plants found in a freshwater pond habitat. This is because animals and plants have certain features and ways of life that are best suited to one particular habitat. For instance, fish have gills which suit them to life in water. Sheep have lungs which suit them to life on land. The features of animals and plants which suit them to the place in which they live are called 'ADAP-TATIONS to their environment'. An important part of any piece of ecological field work is to consider the adaptations to the environment of an animal or plant that is found in a certain habitat.

Community

The animals and plants that live together in a particular habitat are called a community. A community is always changing as different organisms are continually entering and leaving the habitat and as environmental conditions, such as weather, are constantly changing.

There is competition between organisms for food and living space, and the more successful animals and plants eventually replace the less successful. There is a fine balance between the organisms in the community and this balance can be easily upset by humans, disease, POLLUTION or other changes.

Ecosystem (see diagram)

The environment will normally provide organisms with food, shelter and somewhere to reproduce. However, the relationship of organisms to their changing environment is very complicated. The complex relationships between organisms and their environment is called an ecological system or ecosystem.

Energy comes into the ecosystem in the form of heat and light from the sun. This energy is then converted into forms which all living things can use. The green plants can convert the sun's energy into chemical energy by PHOTOSYNTHESIS. Plants are therefore called the PRODUCERS because they make food substances using sunlight. The HERBIVORES get their energy by eating the plants. The CARNIVORES get their energy by eating the herbivores. Herbivores and carnivores are called CONSUMERS. When organisms die, their bodies are decomposed by bacteria and fungi, and this makes the chemicals of their bodies available again for new living things. The bacteria and fungi are called decomposers.

An ecosystem is made up of all the producers, consumers and decomposers in the environment. The circulation in the environment of materials such as water, oxygen, carbon and nitrogen, and the physical part of the environment, such as rocks, soil and air, are also part of the ecosystem.

The relationship between producers, consumers and decomposers in a particular habitat can be shown by making a food web, a food chain or a pyramid of numbers for that habitat. Details are given about these in chapter 8.

The names of animals and plants found in a habitat can be found by using identification KEYS (further details in chapter 1).

Related reading

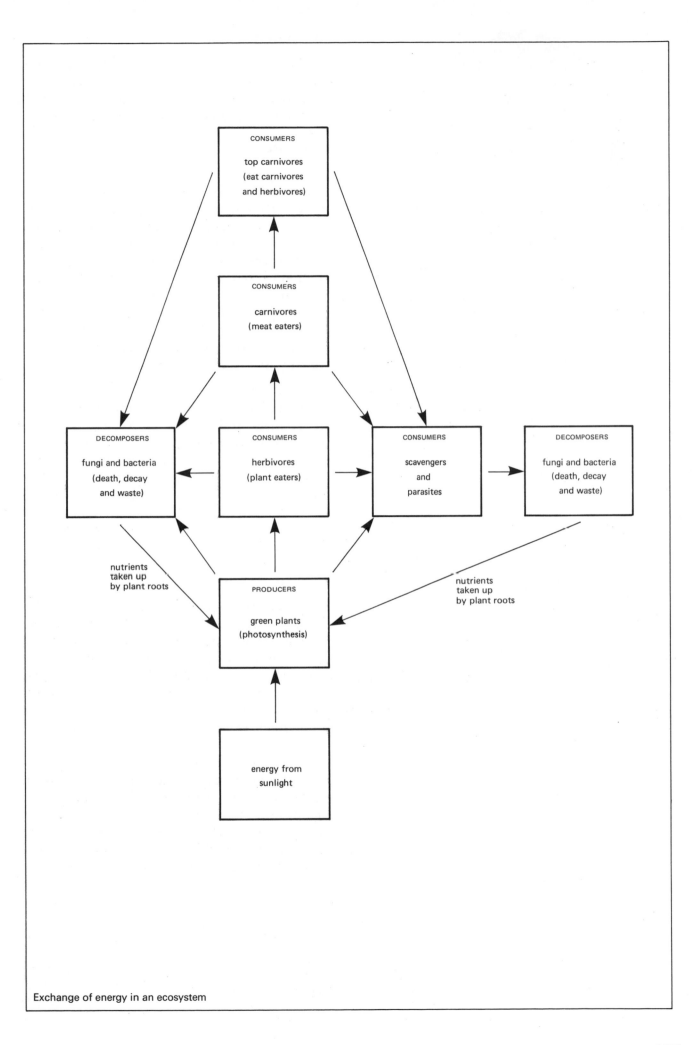

Exchange of energy in an ecosystem

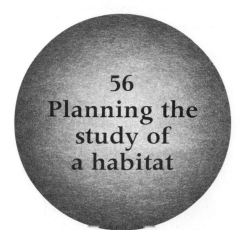

56 Planning the study of a habitat

Ecological fieldwork can be done in many different HABITATS. Information about fresh water, the seashore and soil is given in the next three chapters, but studies can also be made on pathways, hedges, lawns, wasteland, woodland and in many other places. Whichever habitat is chosen for a study, it is important that careful plans are made before beginning.

There are some important practical considerations to be made when deciding on a habitat. Permission may be needed from someone, such as the landowner, before going to a habitat. Also, there may be dangers such as crumbling cliffs, steep slippery banks or badly POLLUTED water.

It is also very important to make sure that the habitat is not disturbed too much by the study. Any rocks or logs which are moved must be put back in the same place otherwise ORGANISMS which live on or under them may die. Identification KEYS should be carried to the area of study so that SPECIES need not be brought back to the laboratory for identification. If a small number are brought to the laboratory for study, then they must be returned to the habitat when the study is finished. Some beautiful habitats have been ruined by careless biology students.

Equipment

A general list of equipment is shown opposite. However, not all of the items will be needed for one particular habitat. For example, the nets will only be needed when studying a water habitat.

Methods

Measuring pH

pH is an indication of the acidity or alkalinity of a liquid. pH paper can be used to test a solution. The paper changes colour in the solution and this colour can then be matched to a colour chart which gives the pH number. pH indicator solution is used in soil test kits. pH 7 means the solution is neutral. pH 8 to pH 14 shows increasing alkalinity. pH 6 to pH 1 shows increasing acidity. The pH of the habitat is important as many species of plants and animals can only live in a small pH range.

A quadrat (see diagram 1)

It would take a long time to count and record all the animals and plants in a large area, so randomly chosen small areas such as a square metre are studied. This can be marked out using pegs and string or a square frame of metal or wood. This is called a quadrat. The various species found in the quadrat can be mapped onto a square drawn on paper using either colours, shapes or shading for the various species.

Quadrats can be made at random places in the habitat or in a line across the habitat.

A transect (see diagram 2)

A line transect can be made by having a long piece of string between two poles and placing it in a straight line across the habitat. A record is then made of everything that the string touches.

A belt transect is similar to a line transect except that everything within $\frac{1}{2}$ metre or 1 metre of the line is recorded.

An ecological study of a habitat should include a description and map of the habitat and a list of the animals and plants found. It should also include an indication of the numbers of animals and the distribution of plants. Some of the species found should then be examined, to find some of their ADAPTATIONS to the ENVIRONMENT.

Related reading

Chapter 55, Ecology
Chapter 57, Fresh water
Chapter 58, Seashore
Chapter 59, Soil

Equipment list

Notebook	Identification keys	pH indicator	Quadrat
Pencils	Specimen tubes	Thermometer	Transect poles
Ruler	Specimen bottles	Light meter	Transect line
Graph paper	Polythene bags	Magnifying lens	Nets
Clip board	Labels	Forceps	White dish
		Knife	Trowel

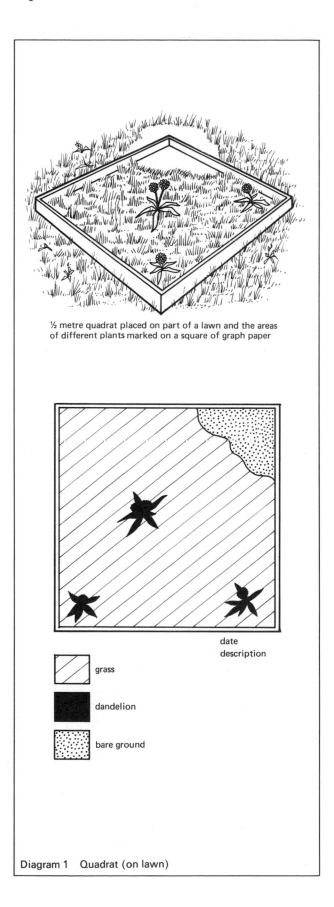

½ metre quadrat placed on part of a lawn and the areas of different plants marked on a square of graph paper

date
description

grass

dandelion

bare ground

Diagram 1   Quadrat (on lawn)

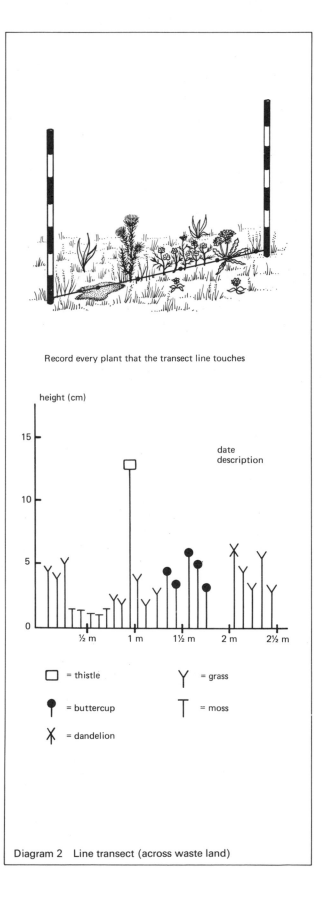

Record every plant that the transect line touches

height (cm)

date
description

= thistle   = grass

= buttercup   = moss

= dandelion

Diagram 2   Line transect (across waste land)

57
Ecology of fresh water

There is a great variety of freshwater HABITATS. They include standing water, in large lakes and small ponds, and running water, in fast-flowing streams and drainage ditches. One type of habitat has different physical and chemical properties from another and these properties determine what SPECIES of animals and plants can live in that habitat. The ecologist must examine these physical and chemical properties so that he can begin to understand why a certain species is present or absent.

Notes should be made about the following properties of the habitat to be studied:

Physical properties

Temperature
Water warms up and cools down much more slowly than air. Because of this, many OR-GANISMS living in water cannot tolerate sudden changes in temperature. Different species survive best at different temperatures.

Light
Water reflects light and so a certain amount of light is lost to submerged plants. The deeper the water, the less light penetrates. Overhanging trees or high banks also reduce the light.

Buoyancy
Water is denser than air so water animals and water plants do not have to support all their weight. They often have a more delicate body structure compared with land animals and plants.

Surface tension
Some organisms, such as water crickets, can support themselves on the surface of standing water. Others, such as water snails and flat-worms, can crawl along the underside of the surface film.

Chemical properties

Oxygen content
The amount of oxygen in the water depends on: (i) the temperature – the warmer the water the less the dissolved oxygen; (ii) the amount of sunlight for PHOTOSYNTHESIS; (iii) the turbulence of the water – when water splashes it mixes with air; (iv) the number of organisms respiring and using up the oxygen. The oxygen content of the water can be measured accurately using chemical analysis or complicated instruments.

pH
The majority of fresh waters have a pH range from 5.5 to 10.0 and this can be measured using indicators. The amount of carbon dioxide and other chemicals such as calcium or nitrates in the water determines the pH, and so the types of organisms living in the water. For instance, 'hard' water contains calcium salts which are needed by the water snails to build their shells.

Further notes should include the date, a map of the habitat and a map reference and details about width, depth, speed of current and whether the bottom is of mud, stones or sand.

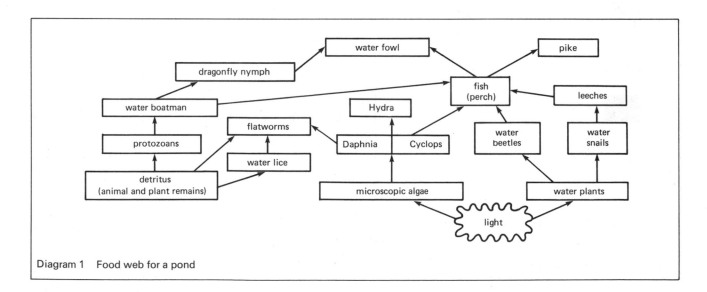

Diagram 1   Food web for a pond

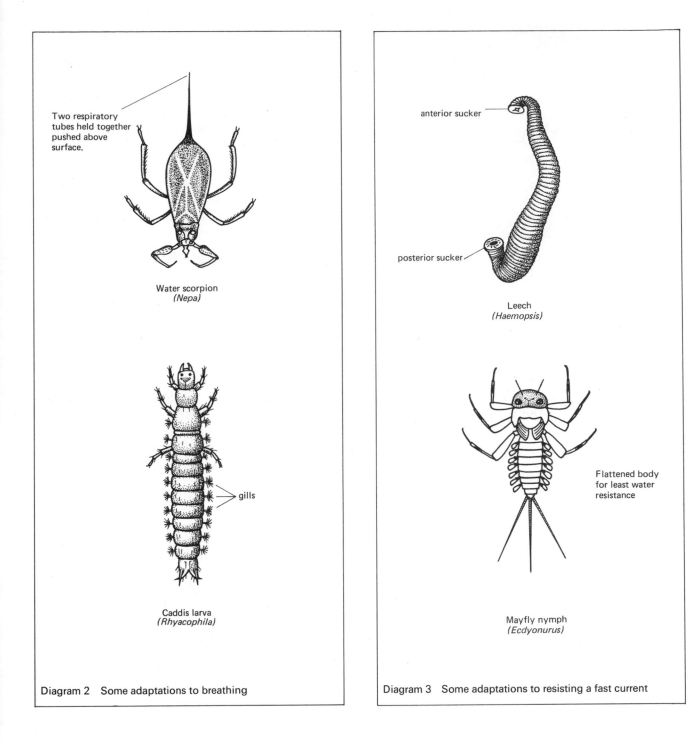

Two respiratory tubes held together pushed above surface.

Water scorpion
(Nepa)

gills

Caddis larva
(Rhyacophila)

anterior sucker

posterior sucker

Leech
(Haemopsis)

Flattened body for least water resistance

Mayfly nymph
(Ecdyonurus)

Diagram 2   Some adaptations to breathing

Diagram 3   Some adaptations to resisting a fast current

Animals and plants

Animal species should be identified and listed according to where they were found. For example: on the surface film; free swimming; attached to a rock or burrowed in the mud. An estimate of the number of each species should also be made. A quadrat will be needed for this purpose. Samples of mud and weeds can be transferred to a white tray or dish and then examined to see if any animals emerge.

Plant species should also be identified and listed according to where they grow. For example, free floating on the surface; rooted in the mud with leaves underwater; by the edge of the water or only covered during floods. A belt transect from one bank to the other may be a useful way of showing the distribution of plants.

When a list of the animal and plant species present has been completed, consideration should be given to the number of each species, their ADAPTATIONS to the ENVIRONMENT (see diagrams 2 and 3) and how they fit into the ecosystem. A food web may be constructed using the species found (see diagram 1).

Related reading

58
Ecology of the seashore

The seashore is formed by the action of the sea and the weather on the land. This may produce a rocky shore with many rock pools, a shingle beach with millions of pebbles, a sandy beach formed from small fine particles of rock, or a muddy shore where the particles are smallest of all. These different seashores will offer different HABITATS for living ORGANISMS. However, all living organisms on a seashore will have to survive the effects of the sea and weather.

Tides

Tides are caused because of the gravitational pull between the earth, moon, sun and planets. Because the earth spins on its axis, the moon rotates around the earth and the earth and moon rotate around the sun, there is a complicated system of tides on earth. However, the effect on living organisms on a seashore is that they must be able to survive being covered by seawater and then exposed to the air for various times each day.

Zonation (see diagrams 2, 3 and 4)

An organism near the upper part of the seashore will spend most of a day out of water. An organism near the lower part of the seashore will spend most of a day covered by water. Thus, on many seashores, particular SPECIES are found at the upper, middle and lower parts of the seashore. A belt transect from the upper shore to the lower shore often shows bands of different species at different areas of the shore. This banding of different species is called ZONATION.

Problems of seashore life

Desiccation (drying up)
Animals and plants must have some way to avoid drying up when the tide is out. For example, seaweeds have slime and barnacles close their shells.

Temperature changes
The temperature of a small rock pool increases quickly during a sunny day at low tide. However, when the tide comes in the pool is rapidly cooled. Living things in that pool must be able to survive these big changes of temperature.

Water regulation
Water evaporates from rock pools as they warm up and so the concentration of salts in them increases. Animals must tolerate corresponding changes in their body fluids to match the concentration of the rock pool.

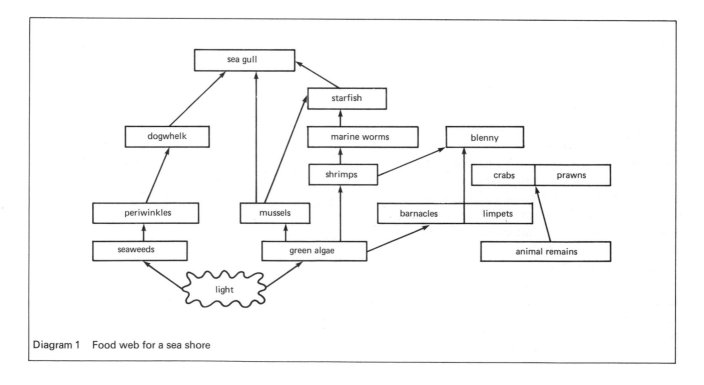

Diagram 1 Food web for a sea shore

Respiration

Only a few seashore animals breathe air. When the tide is out, most seashore animals must be able to manage with less oxygen and survive a high carbon dioxide concentration in their bodies until the tide covers them with water again.

Nutrition

Many seashore animals can feed only in water and so they must be able to feed well at high tide so that they can survive during low tide.

Attachment

Wave action can be very strong and so animals and plants must be able to stop themselves from being washed away. Seaweeds have holdfasts and mussels anchor themselves by threads.

Animals and plants

All animals and plants found on the seashore should be studied carefully for their ADAPTATIONS to the ENVIRONMENT and for ways in which they survive the problems of seashore life. It is important also to note the zonation of the animal or plant to see if it is distributed all over the seashore or in a particular ZONE. A line or belt transect is the best way to discover this.

Rock pools can be mapped and studied using a quadrat. The study should include the same physical and chemical measurements that were taken for a freshwater habitat (chapter 57).

A food web may be constructed using the species found (see diagram 1).

Related reading

Chapter 55, Ecology
Chapter 56, Planning the study of a habitat
Chapter 57, Chemical and physical properties of water
Chapter 1, Use of keys
Chapter 8, Food chains and food webs

Zonation

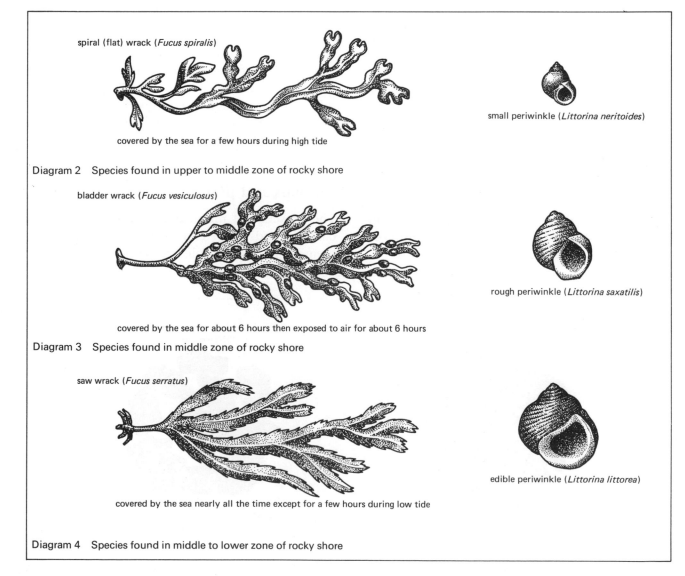

spiral (flat) wrack (*Fucus spiralis*)

covered by the sea for a few hours during high tide

small periwinkle (*Littorina neritoides*)

Diagram 2   Species found in upper to middle zone of rocky shore

bladder wrack (*Fucus vesiculosus*)

covered by the sea for about 6 hours then exposed to air for about 6 hours

rough periwinkle (*Littorina saxatilis*)

Diagram 3   Species found in middle zone of rocky shore

saw wrack (*Fucus serratus*)

edible periwinkle (*Littorina littorea*)

covered by the sea nearly all the time except for a few hours during low tide

Diagram 4   Species found in middle to lower zone of rocky shore

59
A study of soil

Soil is an ENVIRONMENT which provides food and shelter for a large number of living things such as bacteria, fungi, worms and arthropods. Many plants depend on the soil for anchorage, water and minerals.

Soil is mainly a mixture of two types of materials (see diagram 2):

Mineral materials These have come from the break up of rocks by a process called 'weathering'. The size of the material particles and the type of rock from which they came will determine the type of soil.

Humus materials These are formed by the breakdown of animal and plant remains. This breakdown is caused by the bacteria and fungi in the soil. Humus contains the nitrates, phosphates and other salts which are needed by plants for healthy growth. Humus helps to bind the mineral materials in the soil and helps prevent them being blown away by the wind or washed away by water.

Types of soil

There are various types of soil depending upon the mineral materials and the relative amounts of humus, air, water and living things present.

Sandy soils (see diagram 1b)

Sandy soils have large particles which have large air spaces between them. This gives the soil good drainage and AERATION and plant roots can grow through it easily. However, sandy soils do not hold much water and plants quickly die in a drought. The movement of water through the soil can wash away the humus and chemicals needed for healthy growth.

Sandy soils can be improved by adding humus in the form of manure which improves the water-holding and mineral content of the soil.

Clay soils (see diagram 1a)

Clay soils have small particles, small air spaces and poor drainage. Wet clay contains very little air for respiration by animals and by plant roots. Clay soils are often called heavy or sticky because the small particles become packed together into a solid mass. Important chemicals are not washed away as happens in the sandy soils. Clay soils can be improved by digging to break up the lumps, adding humus and sand to improve the drainage and by adding lime which joins the small particles of clay together into large particles. Lime also makes the clay less acid.

Loam soils

Loams are the most fertile soils. They consist of about one half sand, one quarter clay and one quarter humus with a little lime. This forms good soil crumbs which have good drainage and plenty of air spaces but which also hold enough water and minerals for healthy plant growth.

Animals and plants

The presence of microsopic ORGANISMS such as bacteria and fungi can be shown by growing them from a soil sample on a STERILE AGAR plate.

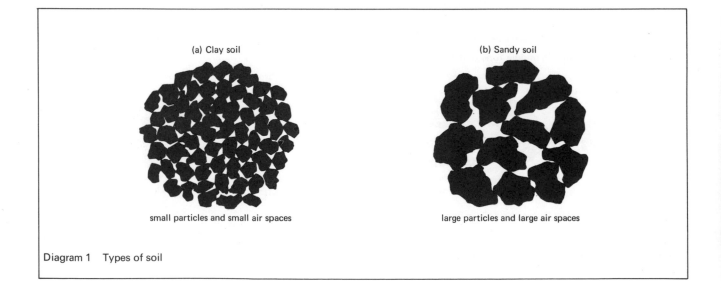

(a) Clay soil

small particles and small air spaces

(b) Sandy soil

large particles and large air spaces

Diagram 1   Types of soil

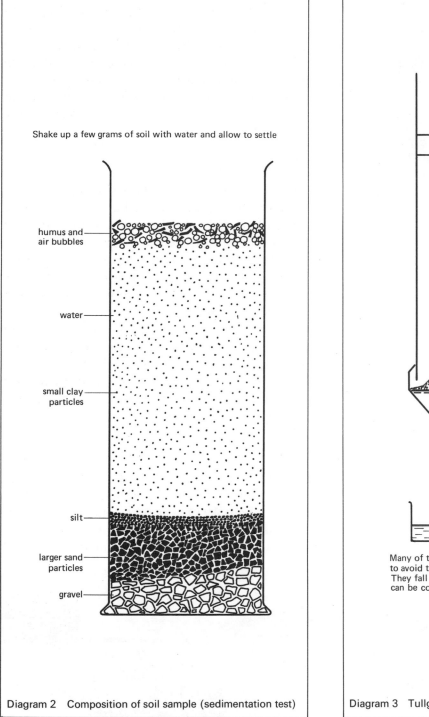

Shake up a few grams of soil with water and allow to settle

humus and air bubbles

water

small clay particles

silt

larger sand particles

gravel

Diagram 2   Composition of soil sample (sedimentation test)

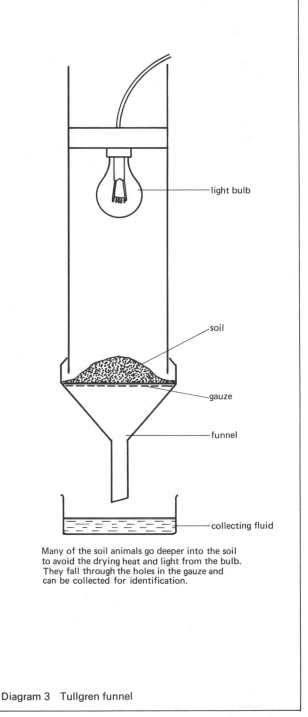

light bulb

soil

gauze

funnel

collecting fluid

Many of the soil animals go deeper into the soil to avoid the drying heat and light from the bulb. They fall through the holes in the gauze and can be collected for identification.

Diagram 3   Tullgren funnel

Small animals such as springtails can be extracted from the soil by using a Tullgren Funnel (see diagram 3) and then identified using a suitable key.

Larger animals, such as snails and insects, on the surface can be studied using a hand lens. Worms can be extracted from the soil by pouring soap solution on the surface. An estimate of the numbers of animals in an area can be made using a quadrat.

Plants only grow well in a soil that suits them. The types of plants growing in a certain area can indicate the pH of the soil. Conifers and rhododendrons grow in an acidic soil whereas fruit trees prefer an alkaline soil. The animals and plants which have been identified should be listed and then studied for their ADAPTATIONS to the environment.

Related reading

60
Man and agriculture

Man as a hunter and gatherer

The SPECIES man has lived on the earth for about one million years. For 30 000 generations (thirty years is one generation) he lived by hunting wild animals and by gathering wild fruits and roots. He developed the use of fire and used simple tools, but he had little effect on his ENVIRONMENT.

Man as a farmer

About 10 000 years ago, man began to collect and plant the seeds from wild grasses to produce crops. These crop-growers developed methods of irrigation, ploughing, harvesting and grinding their seeds. They gradually became settled into communities and had enough food for some people to have leisure time. Leisure time is essential for man to think about inventions and make discoveries.

Farmers began to domesticate and breed animals instead of hunting or herding wild animals. However, this more settled life meant a gradual increase in human population. This then meant more mouths to feed. Trees were cut down to produce more farming land, but some areas were over-grazed by the domestic animals and became desert. Such an area was the Sahara (see diagram 1).

The machine age

The year 1700 A.D. marked the slow beginning of the industrial revolution in Europe. Machinery was made to help the farmers and this changed the pattern of agriculture. The machines were expensive to buy, so over a period of years farms combined, became larger, and specialised in the production of one crop (see photograph). However, this specialisation in farming gives insects, bacteria and fungi an opportunity to grow and reproduce on a large scale. An insect population that lives only on cotton seeds will do very much better on a large specialised cotton farm than on a small mixed farm. Pests have become a problem to farmers since specialisation.

Pesticides

The problem of pests forced man to look for artificial ways of controlling them. DDT was first used in about 1939 as a pest control. However, DDT kills harmless as well as harmful insects and it builds up in the bodies of animals in the food

A specialised farm producing only one crop

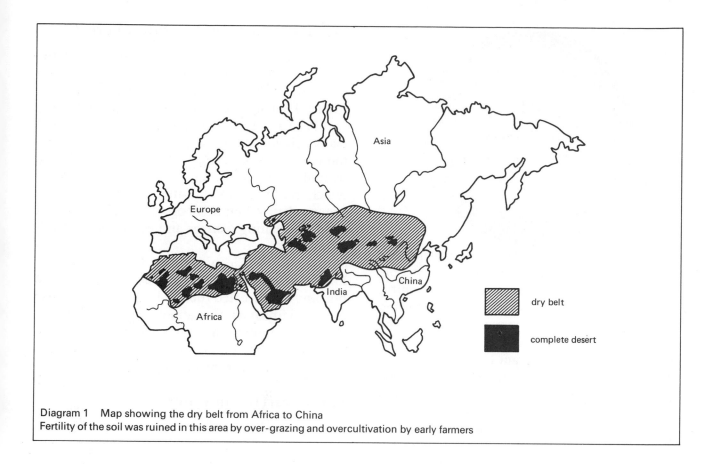

Diagram 1   Map showing the dry belt from Africa to China
Fertility of the soil was ruined in this area by over-grazing and overcultivation by early farmers

chain. This was not understood when DDT was first used, but now insecticides are used which break down quickly into harmless chemicals.

Fertilisers

The amount of food produced can be increased by adding artificial nitrates, phosphates and sulphates to the soil. This is because these chemicals are lost from the soil when crops are gathered. However, fertilisers can be washed through the soil and may get into rivers and lakes which are used for drinking water. The nitrate level in drinking water is now high in some areas of the country.

Weed killers

Weeds are plants that farmers do not want. Early in 1900 man discovered plant HORMONES. These have been developed into selective weed killers as different concentrations are able to kill different types of plants. The widespread use of plant hormones has given man a high crop yield and crops which are easier to harvest.

Summary

Man is having a great effect on his environment in order to feed a rapidly expanding population. He has changed the ecology of large areas of the world. The long-term effects of these changes on the environment are not yet fully understood.

Related reading

Chapter 28, Carbon cycle and nitrogen cycle
Chapter 59, Soil
Chapter 62, Conservation

61
Man and industry

Man's greatest effect on his ENVIRONMENT has always been due to large numbers of people living in a small area. In any animal community, including man, this creates problems of getting enough food and the removal of waste. The world population of man has increased dramatically since 1900 (see diagram 1).

In the late 1700s, steam engines became efficient enough in their use of coal to be used as a source of power. This power was used for pumping water, smelting iron and eventually in the manufacturing industries. From 1800 to the present day, machines have gradually given man an improved standard of living and an increased ability to affect his environment. However, industry created large communities around factories and mines. These large communities have problems getting rid of waste materials.

Waste products in the atmosphere

Burning coal and oil has provided most of the power for industrial societies. Coal and oil are fossil fuels which took millions of years to form from animal and plant remains. These fuels have also been used to make medicines, plastics and many chemicals needed by agriculture and industry. Unless these fuels are scientifically treated while burning, they produce wastes which build up in the atmosphere of a large community. When this POLLUTED air is breathed it can cause lung damage and even lead to death in humans.

In 1956 the 'Clean Air Act' was made law in England and Wales. This act controls the burning of untreated fuels in large towns and has greatly improved the atmospheric conditions. However, motor car exhausts produce very poisonous wastes and as the number of cars increases so the problem is growing.

Wastes carried in water

Man has traditionally got rid of his sewage and solid waste by dumping it in the nearest river or the sea. If one or two households do this, the effect on the environment is not great. But if

Diagram 1 Graph of world population increase since 1600 A.D. (inset shows a graph of numbers in a natural population, see chapter 42.)

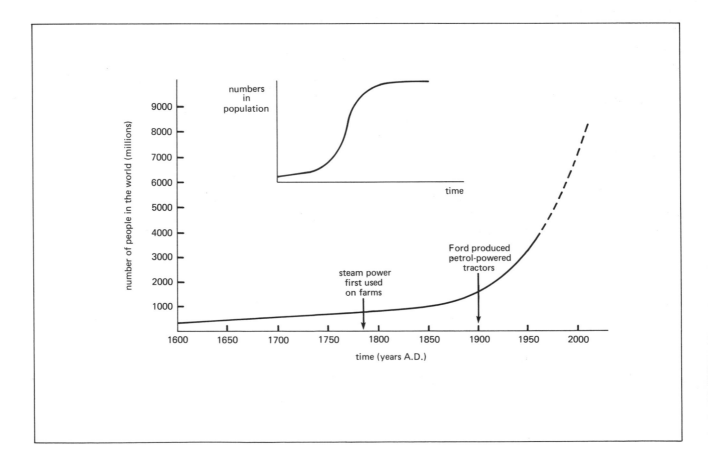

huge amounts of rubbish from large cities and factories are poured into rivers, the plants and animals will soon be affected and many of them will eventually be killed. The bacteria and plants in water would normally turn small amounts of sewage and rubbish into harmless chemicals. But when bacteria and plant life are killed, the water remains polluted and becomes dangerous.

Newer sewage works in London (see diagram 2) and laws to make industry stop dumping waste into the Thames have had a great beneficial effect on the river. Plants and animals are returning to live in the Thames for the first time in centuries.

Atomic energy

Fossil fuels will not last for ever. Industrial man wishes to increase his standard of living, so he must find another fuel. Man first made electricity using atomic power in the 1950s. However, the waste materials from a nuclear reactor remain dangerously radio-active for many years. This waste could so affect the environment as to make it impossible for man to survive at all. Man must find a way of making atomic waste harmless if he is to continue to use atomic power.

Solar energy

Scientists are experimenting with energy from the sun at the moment. Solar energy has no dangerous waste but it is still an expensive method to use.

Summary

Industrial technology has given man a high standard of living and great control over his environment. To continue to enjoy this he must control world population and take great care over the disposal of wastes of all kinds.

Related reading

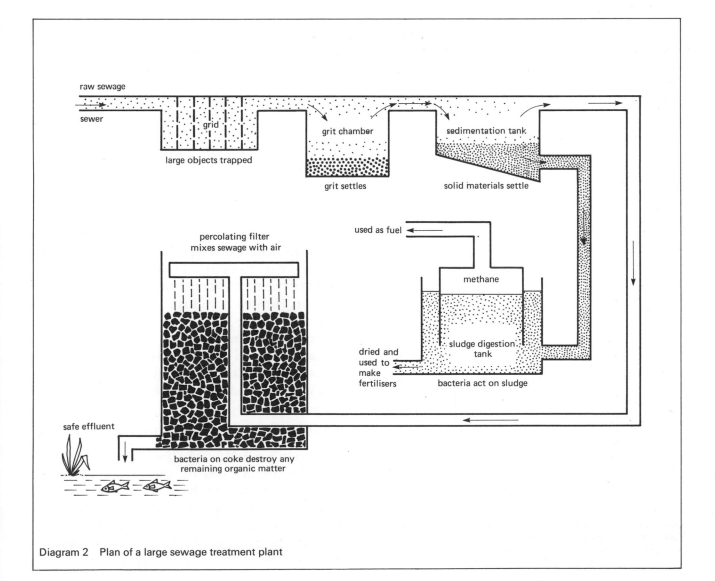

Diagram 2   Plan of a large sewage treatment plant

62 Conservation

Industrial societies enjoy a high standard of living. If this standard is to be extended to the whole of the human population, man must manage the energy supplies and raw materials of the planet more wisely than he has in the past 150 years. Conservation means the wise use of our ENVIRONMENT. Conservation is becoming increasingly important because of the increasing demands for energy and raw materials (see diagram 1).

Conservation of fossil fuels

Coal and oil give heat energy, but they are also valuable sources of complex chemicals. The amount of these fossil fuels on earth is limited and perhaps it would be wiser to use them in making fertilisers, medicines, plastics, dyes, cloth, lubricating oils and many other products rather than just burning them for heat.

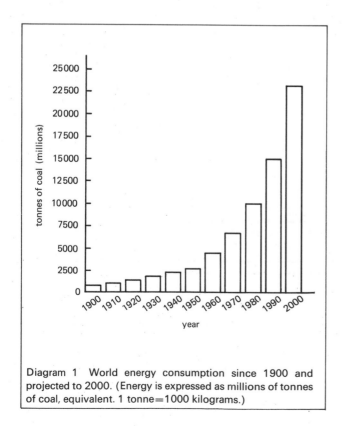

Diagram 1 World energy consumption since 1900 and projected to 2000. (Energy is expressed as millions of tonnes of coal, equivalent. 1 tonne=1000 kilograms.)

If coal and oil are saved for their chemicals, other sources of energy to make heat must replace coke, petrol, diesel oil and fuel oils of all kinds. Most electricity generators are now fuelled by either coal or oil, but man has discovered and developed other techniques of generating electricity.

Alternative sources of energy

Atomic power
This power has been successfully used in England for over twenty years. If it is to be an alternative to fossil fuels, scientists must work on methods of making the waste harmless.

Water power
Water falling from a height can generate electricity but this method is limited to places in the world which are near large rivers and tidal seas. There is no dangerous waste. At present the use of the tide as well as rivers to give a head of water is being investigated by scientists.

Wind power
Windmills have been used by man to provide power for thousands of years. Many conservationists believe wind power could be more widely used.

Solar power
The sun is the main source of power on the planet, but man has only recently learned how to use this power to manufacture electricity. Plants use solar power to make food by PHOTO-SYNTHESIS.

Conservation of raw materials

Minerals, as well as fossil fuels, are man's raw materials. There is only a limited amount of any substance on the earth. Living things constantly RECYCLE chemicals such as nitrogen, oxygen and carbon. However, industrial societies use raw materials faster than they can be naturally recycled. Man must make a deliberate effort to re-use as much of his waste materials as is possible. The minerals in objects such as tin cans and motor cars should be recycled.

Conservation of plant and animal life

Evolution has produced a great variety of SPECIES of living ORGANISMS on the earth. Man must attempt to preserve this variety by not thoughtlessly exterminating any particular species. Wild eagles, giant tortoises, buffalo (see diagram 2) and whales may not seem important to modern man, but the greater variety we can conserve in nature, the greater will be man's variety of food, comfort and enjoyment of his leisure.

Man may find that the removal of organisms which he regards as pests or weeds may produce a result which is not expected. This is because the delicate balance in the environment has been altered.

Summary

Wise use of the environment is essential if man is to continue to enjoy a high standard of living. Wise use requires as many people as possible to learn about man's effect on his environment.

Related reading

Chapter 60, Man and agriculture
Chapter 61, Man and industry
Chapter 28, Raw materials

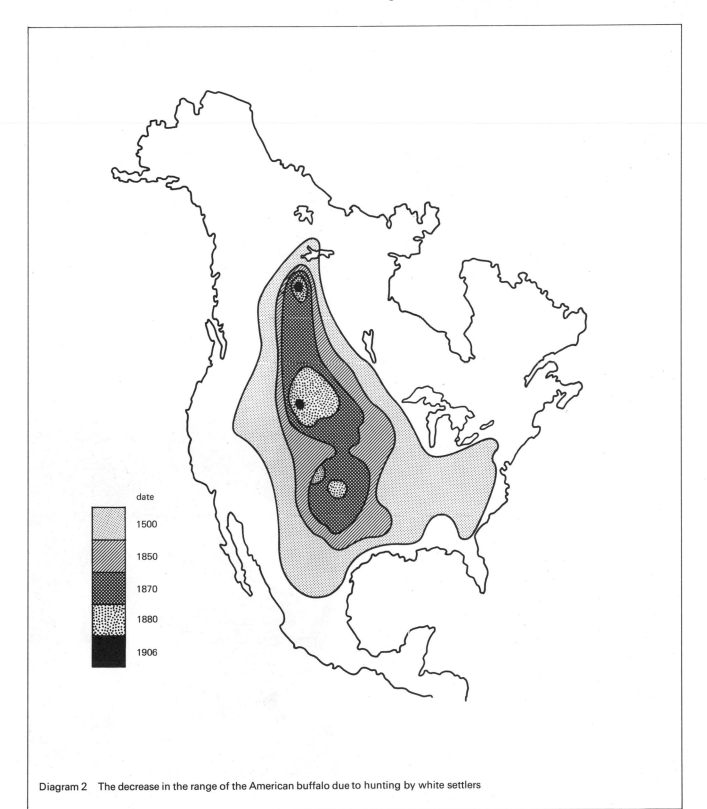

date

1500

1850

1870

1880

1906

Diagram 2   The decrease in the range of the American buffalo due to hunting by white settlers

63
Social animals

If living ORGANISMS are to survive they must have food and space. The weak members of a SPECIES usually fail to obtain these and die. However, the young can be both weak and valuable. The problem of survival of these valuable young has been solved in the animal kingdom by one of two methods. Either vast numbers of young are produced so that chance makes it possible for one or two to survive, or social groups are formed to protect and feed the young while they develop into adults.

Birds and mammals usually form social groups based on a family or group of related families. The social insects form groups which are the offspring of one female, the queen.

The survival value of social groups lies in co-operation and the development of a division of labour. Some individuals are food gatherers, some give the group protection, some look after the young, while others are mainly concerned with breeding. In the social insects these divisions of labour are instinctive but in the higher animals, like mammals, the social behaviour is learned during infancy by imitation of the adults in the group.

Social behaviour in the honey bee

Large numbers of bees live together in a hive. The division of labour within the society is between drones and queens, who breed, and workers, who feed and protect the whole community (see diagrams). All the ways in which bees behave are instinctive. This means that they do not need to learn how to build the brood chambers of the LARVAE, or learn how to feed them or learn how to collect nectar and POLLEN.

Social behaviour in birds

The majority of birds spend most of their lives as part of a flock. Each individual cares only for itself. However when young need to be looked after then male and female parents co-operate in the hatching of eggs and in the defence and feeding of their young. When the young birds can fly and feed themselves all the birds join the flock and care only for themselves.

Social behaviour in mammals

Social behaviour in mammals is concerned with the care and teaching of the young. It is highly developed in wolf-like animals, rats and man. The divisions in the group are usually those of hunters, protectors and those animals who look after the young. The role of any member must be learned by the young through imitating the behaviour of adults in the community.

Many biologists believe that, in rats, the social group extends to all members of a family and that such a family unit will co-operate and share all available food. The members of a family unit will also teach each other about the dangers in the ENVIRONMENT. This teaching is done by complex behavioural signals, such as urinating on poisoned meat. The knowledge that certain commercial rat poisons are bad food has been handed down to many generations of rats thus making the poison far less effective in rat control.

In higher animals, the advantages of social behaviour lie in the efficiency of a division of labour and also in added opportunities to gain and pass on new knowledge to the young.

Man has developed social living to a far greater degree than any other animal. This makes it possible for weak members of a community to survive and perhaps make a valuable contribution to the well-being of the community.

Family life in birds

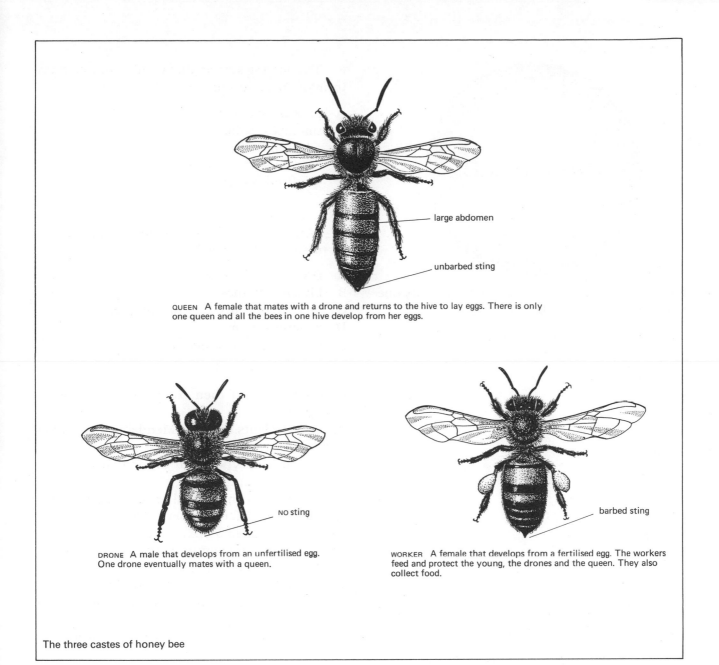

large abdomen

unbarbed sting

QUEEN A female that mates with a drone and returns to the hive to lay eggs. There is only one queen and all the bees in one hive develop from her eggs.

NO sting

DRONE A male that develops from an unfertilised egg. One drone eventually mates with a queen.

barbed sting

WORKER A female that develops from a fertilised egg. The workers feed and protect the young, the drones and the queen. They also collect food.

The three castes of honey bee

Related reading

Chapter 47, Metamorphosis in insects
Chapter 60, Man and agriculture
Chapter 61, Man and industry

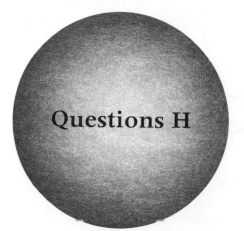

Questions H

The answers to questions 1 to 7 are shown by one of the letters A, B, C, D or E.

1 The genetic material of the nucleus of a gamete is:

A ADP.
B DCPIP.
C TCP.
D DNA.
E ATP.

2 Which one of the following does not have pentadactyl limbs?

A Octopus
B Bat
C Whale
D Crow
E Man

3 Which one of the following statements about water as a habitat for living organisms is FALSE?

A Deep water does not freeze throughout and some organisms can survive below the ice.
B Water helps to support organisms living within it.
C The surface of water focuses light.
D Large volumes of water do not show wide variations in temperature.
E Water contains less oxygen which is available for breathing than does the same volume of air.

4 The effect of adding lime to a soil is to:

A clump the sand grains together.
B release ammonia from manures.
C feed the plant with sodium.
D clump the clay particles together.
E make the soil more acid.

5 The ultimate source of energy for all living organisms is:

A nuclear energy.   D potential energy.
B kinetic energy.   E radioactive energy.
C solar energy.

6 The heating system that would not cause any form of air pollution is:

A paraffin oil heating.
B smokeless solid fuel.
C oil-fired boiler.
D electric-storage heater.
E gas-fired boiler.

7 Which of the following is NOT a characteristic of all social insects?

A Division of labour
B Live in colonies
C Usually produced by one mother
D Beneficial to man
E Behaviour is instinctive

8 Copy out and then complete the following sentences:
(i) The plants used by Mendel in his studies of heredity were..............................
(ii) The genotype of an individual describes ....
(iii) When an individual appears in a species and it is very different from all the others in that species, it is called a ....
(iv) An environment with one particular set of conditions can be called a..............................
(v) A liquid which is pH3 is said to be..........
(vi) A leech is adapted to resisting a fast current by having ....
(vii) One problem of life on the seashore is ....
(viii) The most fertile type of soil is called ....
(ix) The population of the world passed 3000 million in approximately the year....................
(x) Most of the energy for man's industries comes from ....

9 (i) What are chromosomes and where are they found?
(ii) How many chromosomes are there in (a) a normal human body cell and (b) a normal human gamete?
(iii) How is a person's sex determined genetically?
(iv) In humans the allele for brown eyes is dominant to the allele for blue eyes. However, a blue-eyed baby can be produced by a brown eyed father and a brown eyed mother. Explain how this is possible. In your answer use the words homozygous, heterozygous, phenotype, genotype, dominant and recessive.

10 Write about six lines on FIVE of the following:

(i)   Alleles

(ii)   Dominance

(iii)   Fossil records

(iv)   Community

(v)   Use of quadrats

(vi)   Adaptations to respiration in water

(vii)   Zonation on a seashore

(viii) Humus

(ix)   Atomic energy

(x)   The queen bee

11 Write an account about twenty-five lines long on ONE of the following:

(i)   The work of Gregor Mendel

(ii)   Natural selection

(iii)   Physical and chemical properties of water

(iv)   Soil improvement

(v)   Conservation of animal species

12

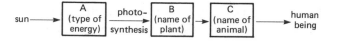

(i) What should be written in the boxes to replace the letters A, B, C, in this energy flow diagram?

(ii) The diet of humans is called omnivorous. What word is used to describe the diet of (a) an animal that eats only plants and (b) an animal that eats only meat?

(iii) Explain briefly the importance in an ecosystem of (a) plants and (b) decomposers.

13 The peppered moth exists in two forms, the pale grey form and the black form. When stationary, it is usually found clinging to the bark of trees. The map of England shows the distribution of the two forms in this country. The proportion of the grey form in a particular area is shown by the white portion of the square and the proportion of the black form is shown by the black portion of the square.

(i)   Explain why only the grey form is found in Devon (D), North Wales (W), and the Isle of Wight (I).

(ii)   Explain why only the black form is found in the Sheffield area (S) and on Canvey Island (C).

(iii)   Why is the grey form present to some extent in London (L) and Manchester (M)?

(iv)   Suggest why area E should have such a high proportion of the black form.

(v)   Until about 150 years ago the black form was very rare indeed anywhere in England. What biological reason is there to account for the rise in proportion of the black moths?

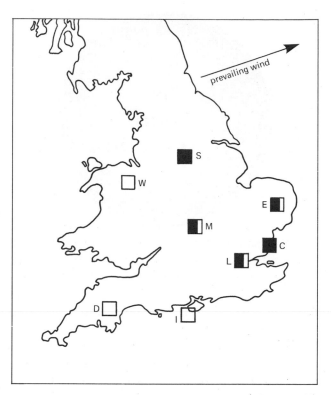

(Met. REB)

14 Soil is formed by the weathering of rocks.

(i)   Describe two different ways in which weathering can occur.

(ii) Describe an apparatus and a method to extract small animals from a soil sample.

(iii) Name one animal found in soil and describe how it is adapted to its environment.

(iv) Describe a clay soil and the ways in which a gardener can improve it.

15 (i) Name a habitat you have studied.

(ii) Describe the method you used to estimate the numbers of a certain species in that habitat.

(iii) Name three animals you found.

(iv) Describe how one of these animals is adapted to its environment.

(v) Name two plants you found.

(vi) Draw a diagram of one of these plants and label on the diagram any features which adapt the plant to its environment.

64
Food preservation

For thousands of years man has preserved his food in times of plenty so that he would not go hungry in times of famine. Industrial man preserves his food for this practical reason as well as for ease of transportation, increased variety and pleasure in eating.

Food is animal and plant material which is usually dead. All such material is made of cells and can go bad for two reasons. Either the cells are attacked and eaten by bacteria and fungi, or the natural ENZYMES inside all living cells begin to break down the cell from inside. Any successful method of preservation must prevent these two processes from taking place.

Bacteria and fungi are living ORGANISMS and can be killed by heat. They can also be prevented from reproducing by very low temperatures. Enzymes are biological chemicals, their structure can be changed by heat and their reaction rate can be slowed down by low temperatures. From this it can be seen that very high and very low temperatures help in the preservation of food.

Bacteria are all around us in air, soil, water and on our bodies. Therefore, once bacteria in food have been killed, the food must be totally protected from any further contamination by bacteria in the surroundings. Many early methods of preservation completely changed the taste of the food preserved. Modern methods attempt to keep the taste and texture of the fresh food as far as possible.

Methods of preservation

Curing

Fish and meat can be preserved by smoking and salting. Both these methods kill bacteria by chemicals and by drying out the food. However, the taste of the food is changed. For example smoked herrings, called kippers, taste different from fresh herrings.

Pickling

Bacteria cannot survive acid conditions therefore food placed in vinegar will not go bad. Examples of this method of preservation are pickled onions and pickled eggs.

Canning

This is a method of food preservation which uses heat to kill bacteria and destroy enzymes. The food is sealed in a container, usually metal, and then pressure-cooked (see diagram). The sealed container protects the food from contamination by bacteria after cooking. Metal cans are usually lacquered inside to prevent the food dissolving the metal container.

Dehydration

Bacteria, like all living things, cannot grow without water. Dried food can be kept safely for long periods at normal temperatures, but these foods should have water returned to them before being eaten.

Refrigeration

Both the action of enzymes and the growth of bacteria are slowed down by temperatures below the freezing point of water. The bacteria are not necessarily killed, their activities are merely slowed down. Very low temperatures, such as those of a deep-freeze unit, can preserve food for years. However the texture of soft fruits, and even meat and fish, is affected unless they are frozen very rapidly. Even then the cells of soft fruits like strawberries are broken by ice crystals inside the fruit. Rapid freezing helps to retain the taste and texture of foods because the ice crystals formed inside cells are very small. Rapid thawing also helps keep texture.

It is important to note that once canned, dehydrated or refrigerated food is returned to normal room temperatures, exposed to air and handled by people, it will go bad just as quickly as fresh food.

Related reading

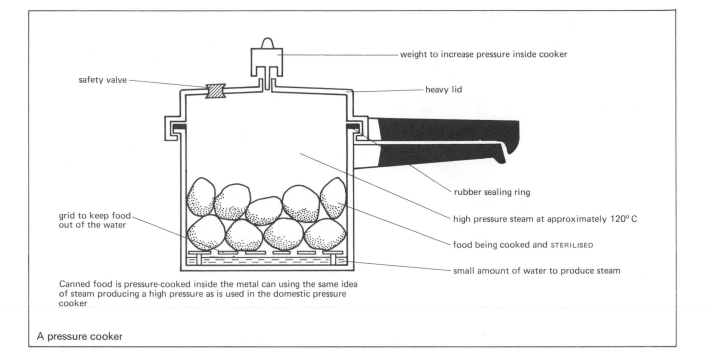

weight to increase pressure inside cooker

safety valve

heavy lid

rubber sealing ring

grid to keep food out of the water

high pressure steam at approximately 120° C

food being cooked and STERILISED

small amount of water to produce steam

Canned food is pressure-cooked inside the metal can using the same idea of steam producing a high pressure as is used in the domestic pressure cooker

A pressure cooker

A canning factory

65
Agents of disease

All living things eventually die. Two common causes of death are old age and disease. Also disease can cause weakening of an ORGANISM but, if the cause of disease is removed, the organism can recover good health.

Causes of disease

There are two main causes of disease:
Lack of essential raw materials. For example, lack of vitamin C in humans causes scurvy.
Invasion of the body by other living organisms which deprive the HOST of food and also make poisonous chemicals which interfere with the working of the host's body.

Invasion can be by many types of organisms:
(i) Many-celled parasites, for example, tapeworm in humans and liver fluke in sheep.
(ii) Fungal infections, for example, ringworm and athlete's foot in humans and bracket fungus in birch trees.
(iii) One-celled animals, for example, Plasmodium, living in the blood and liver of humans cause malaria.

(iv) One-celled organisms called bacteria. Most bacteria are completely harmless. However some invade larger organisms and produce disease, for example, venereal diseases, pneumonia, tuberculosis, cholera and tetanus in humans.
(v) Viruses. These cause a large number of diseases in all living organisms, for example, poliomyelitis, smallpox and influenza in humans.

Bacteria (see diagram 1)

These are very small organisms and can be seen only under high magnification. This makes it difficult to detect their presence unless they are very numerous. However, they reproduce very rapidly when living conditions are in their favour. In these conditions a single bacterium will reproduce fast enough to produce a clump visible to the naked eye in twenty-four hours.

Bacteria can be grown, or 'cultured', in laboratory conditions. Experiments are done in hospitals to find which bacterium causes a particular disease.

Culturing bacteria (see diagram 2)
Bacteria need food, moisture and warmth to grow. This can be provided outside a living body by specially-made nutrient broth. These broths can be turned into a jelly by adding a seaweed extract called AGAR.

All materials and apparatus used to investigate bacteria must be carefully STERILISED before the investigation, and a CONTROL must always be used. Bacteria can be dangerous in certain circumstances. Great care must always be taken when handling cultures.

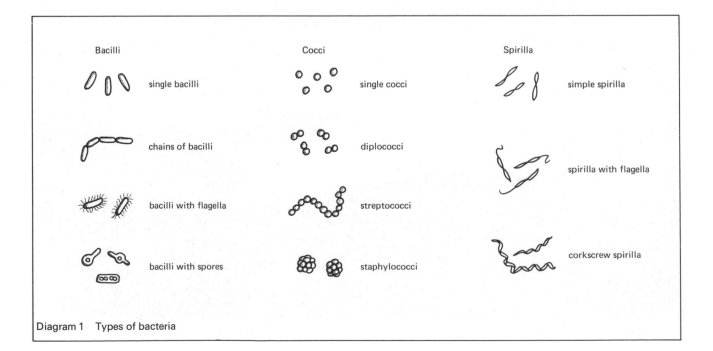

Diagram 1   Types of bacteria

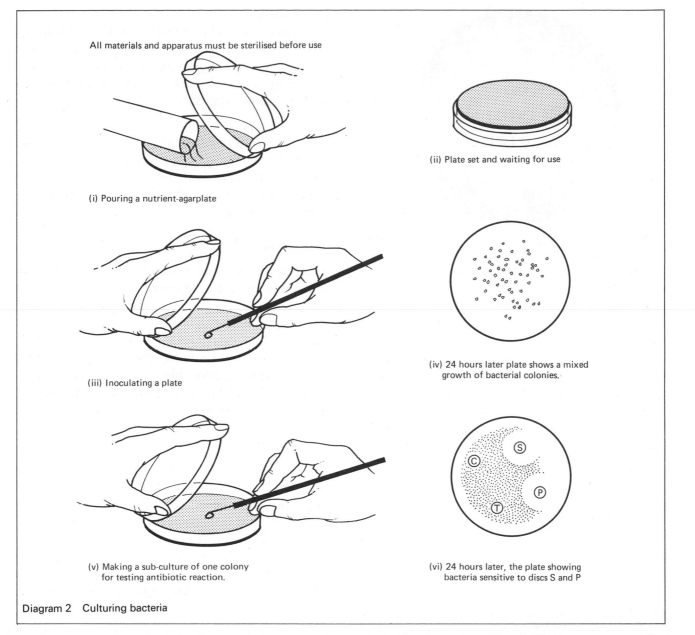

All materials and apparatus must be sterilised before use

(i) Pouring a nutrient-agarplate

(ii) Plate set and waiting for use

(iii) Inoculating a plate

(iv) 24 hours later plate shows a mixed growth of bacterial colonies.

(v) Making a sub-culture of one colony for testing antibiotic reaction.

(vi) 24 hours later, the plate showing bacteria sensitive to discs S and P

Diagram 2   Culturing bacteria

The nutrient agar is melted, poured into a petri dish and allowed to set. The dish is then exposed for a short time to a place where bacteria are suspected to be.

The plate is then closed and placed in an INCUBATOR for twenty-four hours at 37°C, body temperature.

If bacteria are present they will grow in clumps or colonies. When the bacteria have grown, one type can be chosen, removed from the plate and grown alone on a new plate. This is called a sub-culture. Hospitals use discs soaked in different antibiotics, which are placed on the sub-culture plates and incubated. This can show the power of these antibiotics to kill or slow down the rate of growth of the bacteria.

Antibiotics

Certain chemicals, first found in the fungus *Penicillium*, can interfere with the ability of cells to make new protein. When a many-celled animal takes these chemicals the effect is spread over all the cells of the body and the animal can survive. Bacteria are single-celled organisms which cannot survive the chemical interruption to their growth and so many die.

Man has produced many different types of antibiotics to fight the many types of bacteria which produce disease.

Related reading

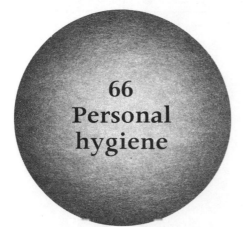

66
Personal hygiene

Hygiene is the science of maintaining good health. The public health of most industrial societies is looked after by the community. Clean water, clean air, removal of sewage and satisfactory housing conditions all help maintain good health. However, without personal hygiene, humans will still suffer from the effects of dirt. Dirt has always been associated with disease even before scientists discovered bacteria. Personal hygiene means keeping one's own body clean and healthy (see diagram).

Care of the skin

Bacteria are all around us and it is impossible to avoid having bacteria on the skin. However, disease only results when the bacteria penetrate the outer layer either through a cut or by way of the sweat glands and hair follicles. Boils are caused by bacteria called staphylococci entering the hair follicles. The bacteria on the skin surface can be kept to small numbers by regular washing. Soap and detergents dissolve the oils and fats which help bacteria to stick to the skin surface and so remove many of them. The hands, which are used for so many tasks such as eating and working, should be washed many times a day, especially after each visit to the lavatory.

Body odour (B.O.)
Healthy humans sweat most of the time, but especially in hot conditions. Sweat is excellent for bacterial growth and so bacteria can increase their numbers rapidly in areas of heavy sweating like the armpits and groin. The waste products of the bacteria build up in these areas and an unpleasant smell is produced if the sweat is not removed at least once a day. Fresh sweat does not have a bad smell. Washing is a more healthy way of preventing B.O. than using deodorants as these products can cause irritation and block sweat glands.

Clean clothes
Many bacteria and PARASITES, some of which can carry diseases, live in clothing as well as on the skin. Regular changing and washing of clothes, especially underclothes, is very important for good personal hygiene.

Small wounds
Any breaks in the skin allow bacteria to invade the body. These breaks should be quickly and thoroughly cleaned. If the break is more than a very small wound it should be covered with a STERILE dressing.

The nose

Bacteria and viruses travel in the air and air is taken into the lungs at every breath. The nose is lined with hairs, which trap large particles of dirt and dust, and with MUCUS which traps many bacteria. The nasal passages should be kept clear by occasionally blowing the nose, preferably into a clean disposable paper handkerchief. Breathing should be through the nose on as many occasions as possible.

The mouth

The mouth is used for talking, eating and occasionally breathing, therefore bacteria can enter and can reproduce causing tooth decay in the mouth and infections of the throat and lungs. Regular cleaning of teeth can remove bacteria from around the teeth, but only good general health can help the body resist infections such as bronchitis, tonsilitis, laryngitis and the common cold.

Summary

Personal hygiene is keeping the body healthy. As well as cleanliness, human beings need a balanced diet, sensible exercise and enough sleep to maintain good health and fight disease.

Related reading

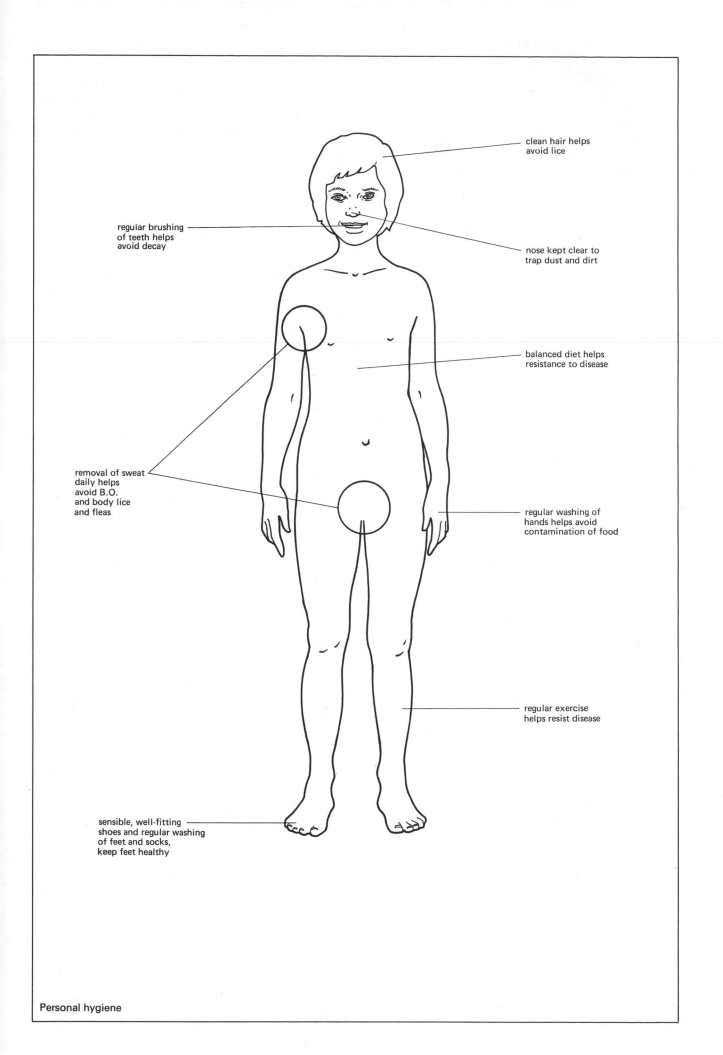

clean hair helps
avoid lice

regular brushing
of teeth helps
avoid decay

nose kept clear to
trap dust and dirt

balanced diet helps
resistance to disease

removal of sweat
daily helps
avoid B.O.
and body lice
and fleas

regular washing of
hands helps avoid
contamination of food

regular exercise
helps resist disease

sensible, well-fitting
shoes and regular washing
of feet and socks,
keep feet healthy

Personal hygiene

67
Parasites

PARASITES are animals or plants which feed on living organic material. The ORGANISM on or in which a parasite feeds is called the parasite's HOST. Most parasites do not kill the host but damage it, causing disease and weakness.

Some parasites are external. For example, fleas and lice live in the hair of mammals and suck their blood. Aphids, such as greenfly, live on plants and take up the sap from the plant. Other parasites are internal. Examples of these are tapeworms, liver flukes, hookworms, malaria parasites and some fungi.

Plant parasites

Most plant parasites are bacteria or fungi. However, there are a few flowering plants which are parasites. Mistletoe is a parasitic flowering plant which depends on its host, usually apple or poplar, for water and minerals. However, mistletoe can make its own food in its green leaves by PHOTOSYNTHESIS. Dodder (see diagram 1) is a parasitic plant which has no green leaves or

roots, so it is totally dependent on its host. The stem of dodder looks like a strand of pink cotton and coils around its host plant, which can be the nettle, clover, gorse or heather. Small suckers grow from the pink strands into the host stem where they enter the VASCULAR BUNDLES and absorb food and water. Dodder produces tiny pink flowers which develop into seeds.

Animal parasites

An example of an animal parasite is the pork tapeworm, *Taenia solium*, (see diagram 2). This tapeworm lives in the intestine of man and so man is called the primary host. But the eggs must spend some time in a pig, so the pig is called the secondary host.

The tapeworm has a small head which is attached to the wall of the intestine of the host by hooks and suckers. Behind the head grows a large number of segments. Each segment absorbs food through its walls from the already digested food in the intestine of the host. Each segment contains male and female reproductive ORGANS which produce eggs. All parasites which have two hosts must produce large numbers of eggs as only a few will reach the secondary host.

Tapeworm infection in humans is rare in most developed countries because (i) untreated human FAECES is not used as fertiliser, so the eggs cannot find a secondary host, (ii) thorough cooking of meat kills any bladderworms and (iii) the standards of food handling and of hygiene are high.

Symbiosis

When two organisms live together it is not always as parasite and host. Sometimes both organisms benefit from the partnership. For example, the nitrogen-fixing bacteria on the roots of some plants, or bacteria in the digestive system of some animals. When two organisms live together and both benefit from the partnership, this is called symbiosis.

Saprophytes

SAPROPHYTES are different from parasites because they are plants which feed on dead organic material. Many bacteria and some of the fungi, such as yeast and mucor, are saprophytes.

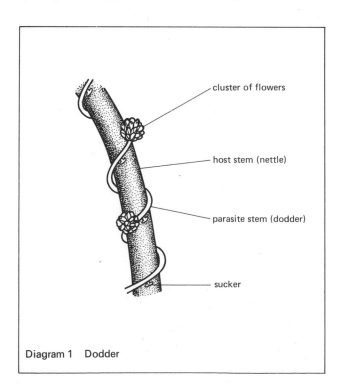

Diagram 1   Dodder

- cluster of flowers
- host stem (nettle)
- parasite stem (dodder)
- sucker

Related reading

Chapter 8, Parasites and saprophytes

158

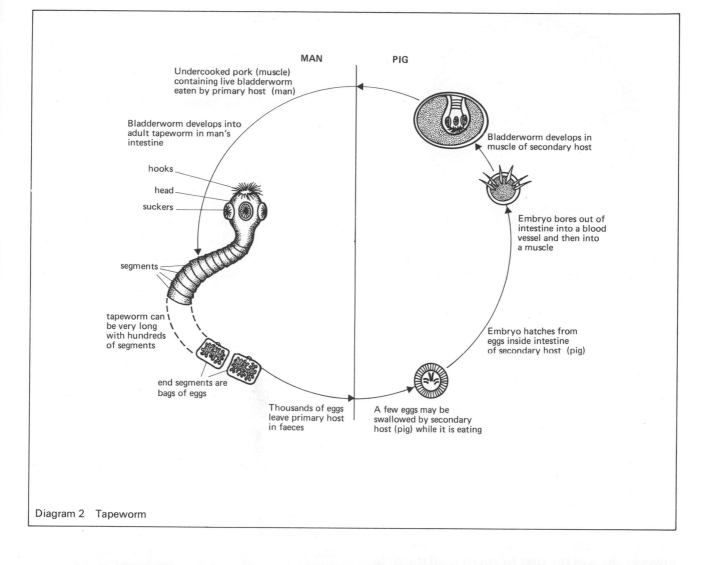

MAN | PIG

Undercooked pork (muscle) containing live bladderworm eaten by primary host (man)

Bladderworm develops into adult tapeworm in man's intestine

hooks

head

suckers

segments

tapeworm can be very long with hundreds of segments

end segments are bags of eggs

Thousands of eggs leave primary host in faeces

Bladderworm develops in muscle of secondary host

Embryo bores out of intestine into a blood vessel and then into a muscle

Embryo hatches from eggs inside intestine of secondary host (pig)

A few eggs may be swallowed by secondary host (pig) while it is eating

Diagram 2   Tapeworm

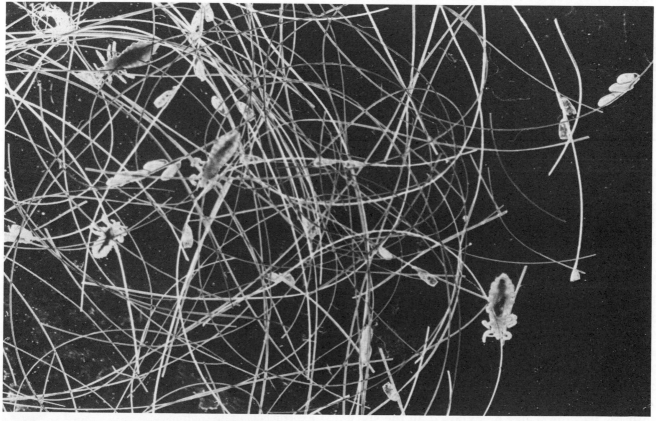

An external parasite — the head louse and its eggs (nits)

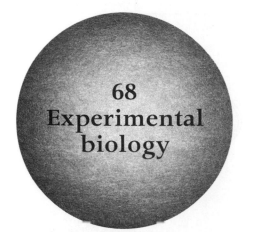

68
Experimental biology

Most people take notice of the world around them. They notice that plants start to grow in the spring and that leaves fall in the autumn. Some of these people may wonder why these things happen and would look for their answer in a book. However, the people who wrote the books and scientists who are asking questions, the answers for which cannot be found in books, must find out by designing and doing experiments.

Observations and hypotheses

Nearly all biological experiments start by someone observing what is happening around them and asking questions. Sir Alexander Fleming discovered penicillin, but was probably not the first man to observe that the fungus *Penicillium* destroyed cultures of bacteria. However, he was the first to ask himself the right questions about *Penicillium* and to suggest possible answers. He then designed and did experiments to try to find out whether his answers to the questions were correct. This led to the discovery of an antibiotic called penicillin. Other biologists continued his work and this has led to the widespread use of various antibiotics in present-day medicine.

An intelligent, scientific guess at an answer to a question is called a hypothesis. An observation is the first stage of a biological investigation; making a hypothesis is the second stage and designing and doing experiments to test the hypothesis is the third stage.

Controlled experiments on living things

The third stage of a biological investigation is to design experiments which will test a hypothesis.

Louis Pasteur, one of the greatest experimental biologists, designed an experiment in 1881 to test a vaccine which he believed would protect cattle against the disease anthrax. He collected together sixty animals, half of which he injected with his vaccine. Later he exposed all the sixty animals to anthrax germs. He found a few days later that the thirty vaccinated animals were well, but the thirty unvaccinated animals had died of anthrax. The conclusion that Pasteur made from this experiment was that the vaccine was successful in protecting the animals. The experiment shows two important facts about biological investigations. Firstly, Pasteur used sixty animals, not just one. Living things are very complicated and even members of the same SPECIES have many differences. This variation must be considered in experiments. If Pasteur had used only one animal, it may have been possible that it had some natural resistance to the disease. Biological experiments on living things have to use many ORGANISMS and be repeated many times to be certain of the results.

Secondly, Pasteur only injected half of the animals with vaccine. The other half were called his CONTROL group. A control group or control experiment must be identical to the actual experiment except for just one factor. Then, if there is a difference in the results between the two experiments, it must be due to that one factor. Pasteur was able to conclude that his vaccine protected the injected animals because his control group all died. If he had originally injected all sixty animals he could not have been so sure.

Controls are needed in most biology experiments in order to be sure that it is the one factor under test that is giving the results in the experiment. The two examples opposite show the importance of controls (see diagrams 1 and 2).

Related reading

Chapter 70, Louis Pasteur
Chapters 6, 12, 14, 15, 16 and 51,
Experiments using controls

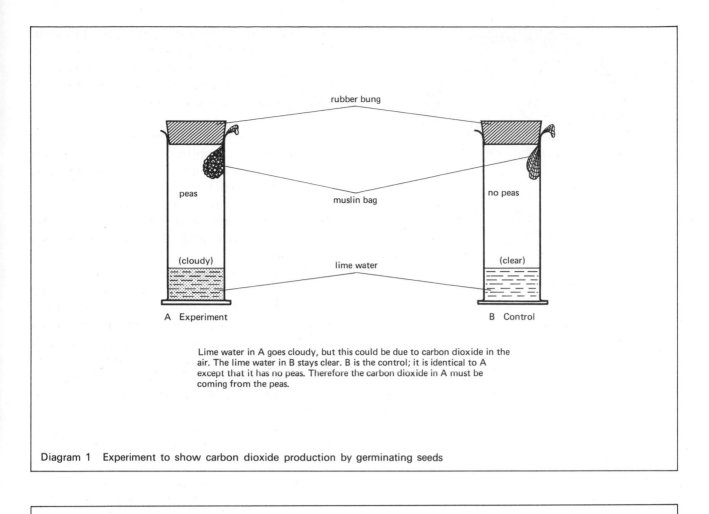

Lime water in A goes cloudy, but this could be due to carbon dioxide in the air. The lime water in B stays clear. B is the control; it is identical to A except that it has no peas. Therefore the carbon dioxide in A must be coming from the peas.

Diagram 1   Experiment to show carbon dioxide production by germinating seeds

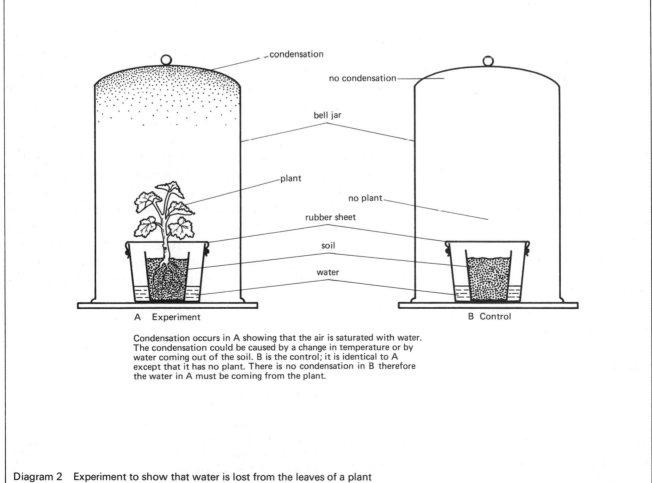

Condensation occurs in A showing that the air is saturated with water. The condensation could be caused by a change in temperature or by water coming out of the soil. B is the control; it is identical to A except that it has no plant. There is no condensation in B therefore the water in A must be coming from the plant.

Diagram 2   Experiment to show that water is lost from the leaves of a plant

69
History of biology

Biology is the science in which living things are studied. All science gains information by observation of the world, by putting forward ideas which might explain the observations and then by testing the ideas with experiments. A new idea may be tested by many different scientists and, if the idea is supported by most of them, it becomes regarded as scientific fact or truth. A scientific idea is often called a hypothesis. In science, 'truth' can be defined as a well-supported hypothesis.

Testing the truth of an idea requires good 'tools of the trade'. Biologists need to be able to observe the very small and so must have good microscopes. Biology also needs good communications between scientists of all countries and languages. This needs a naming system, or classification, of living things which is understood by all biologists. Both these 'tools' became available to biologists in the seventeenth century.

Microscopy

Dutch lens grinders produced the earliest microscopes in the mid-1600s. Before this, in 1616, Harvey (1578–1657), an English physician, had suggested the pathway of the blood around the body. However, it was in 1661, four years after Harvey's death, that Malpighi, an Italian physician, actually saw CAPILLARIES under the microscope and this supported Harvey's original hypothesis. Robert Hooke observed and named plant cells in 1660 and Anton van Leeuwenhoek (1632–1723) was the first man to see bacteria.

Classification

As long ago as 350 B.C., Aristotle observed and classified a number of animals and plants. However, by the seventeenth century this system was not satisfactory. In the 1660s, an English naturalist, John Ray, defined a SPECIES as ORGANISMS similar enough to breed together.

This work was followed by Karl Linnaeus (1707–1778). Linnaeus gave his name to a system of classification still used today, and made it possible for biologists all over the world to communicate with each other.

The development of microscopy and classification in the seventeenth century made possible the advances in biology of the eighteenth and nineteenth centuries.

In 1796, Ingen-Houz described the relationship between light, carbon dioxide and oxygen in plants and thus first described PHOTOSYNTHESIS.

Louis Pasteur (1822–1895) disproved the theory of spontaneous generation and explained fermentation. Charles Darwin (1809–1882) published his book on a theory of evolution, called 'The Origin of Species' in 1859. Gregor Mendel (1822–1884) suggested a hypothesis of the mechanism of heredity based on work done on garden peas. In 1895, Wilhelm Röntgen discovered X-rays.

The twentieth century has seen a great advance in biological knowledge, mostly through co-operation between biology, chemistry and physics. The electron microscope and the development in the 1950s of radioactive tracers made it possible to explain the STRUCTURE and FUNCTION of living systems inside the cell. Perhaps the most exciting theory of the twentieth century so far has been by Crick and Watson in 1953 who described a structure for DNA. DNA is the material of inheritance inside the NUCLEUS of all living cells. This structure has been found to be the same in nearly all animals and plants.

A new area of importance in biology has emerged in the mid-twentieth century. This is the study of living things in their natural ENVIRONMENT called ecology. Ecology describes the relationships in an environment and can be used to predict future events in that environment. Biology, through agriculture, medicine and ecology, is emerging as a study vital to man's future survival on the earth.

Related reading

Summary of important dates in the history of biology

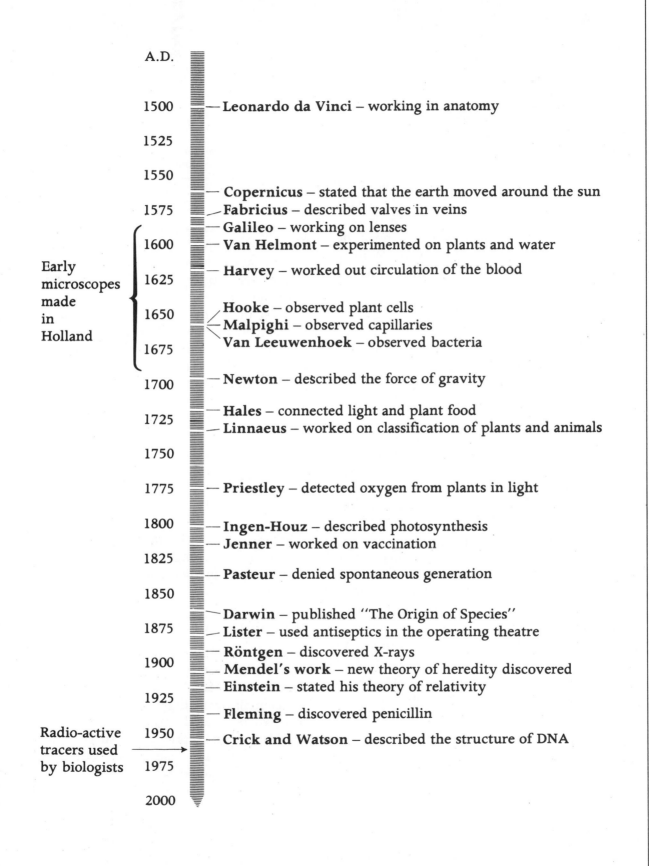

A.D.	
1500	— **Leonardo da Vinci** – working in anatomy
1525	
1550	
	— **Copernicus** – stated that the earth moved around the sun
1575	— **Fabricius** – described valves in veins
	— **Galileo** – working on lenses
1600	— **Van Helmont** – experimented on plants and water
1625	— **Harvey** – worked out circulation of the blood
1650	**Hooke** – observed plant cells
	Malpighi – observed capillaries
1675	**Van Leeuwenhoek** – observed bacteria
1700	— **Newton** – described the force of gravity
1725	— **Hales** – connected light and plant food
	— **Linnaeus** – worked on classification of plants and animals
1750	
1775	— **Priestley** – detected oxygen from plants in light
1800	— **Ingen-Houz** – described photosynthesis
	— **Jenner** – worked on vaccination
1825	
	— **Pasteur** – denied spontaneous generation
1850	
	— **Darwin** – published "The Origin of Species"
1875	— **Lister** – used antiseptics in the operating theatre
	— **Röntgen** – discovered X-rays
1900	— **Mendel's work** – new theory of heredity discovered
	— **Einstein** – stated his theory of relativity
1925	
	— **Fleming** – discovered penicillin
1950	— **Crick and Watson** – described the structure of DNA
1975	
2000	

Early microscopes made in Holland (bracket spanning c. 1590–1690)

Radio-active tracers used by biologists → (arrow pointing to c. 1950)

70
Some famous biologists

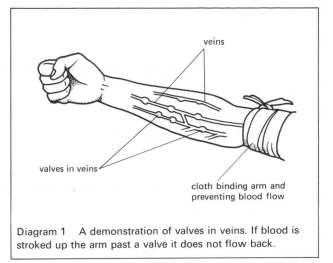

Diagram 1 A demonstration of valves in veins. If blood is stroked up the arm past a valve it does not flow back.

In the early days of science in Europe, biology and medicine were often closely linked. Many of the ideas that were to help the understanding of living things began as ideas to help man's understanding of himself. One idea which gave biology a basis for accurate understanding of the workings of animals' bodies was William Harvey's description of the circulation of blood.

William Harvey (1578–1657)

Harvey was born in Folkestone and went to Cambridge University in 1584. At this time it was believed that arteries contained air, not blood, and that blood ebbed and flowed about the body in veins. Later, Harvey worked as a doctor and teacher at St Bartholomew's Hospital in London. In 1616 he gave a lecture where he stated his hypothesis about blood flowing around the body in a continuous system of VESSELS. His evidence to support his idea was:

1 Veins have valves to stop blood flowing away from the heart.
2 The amount of blood pumped out by the heart every twenty-four hours is far too much to be all newly-made by the body in that time.
3 Pressure on an artery shows that it becomes empty on the side furthest away from the heart. The opposite is true for veins. (See diagram 1.)
Harvey could not actually see the pathway of blood from arteries to veins because he had no microscope with which to see CAPILLARIES. However, his evidence was accepted by most doctors of his day.

Many years after Harvey, another great biological idea was suggested by Louis Pasteur, a French chemist.

Louis Pasteur (1822–1895)

Pasteur was born in Paris and became Professor of Chemistry at the Sorbonne in 1867. At this time it was believed that living ORGANISMS could arise spontaneously from dead material. Pasteur, through his work on wine and fermentation, had discovered a great deal about bacteria and fungi. He set out to demonstrate that living organisms could arise only from living material. (See diagram 2.)

First of all he placed broth in several containers. He then boiled the broth to kill all living organisms. The neck of the flask containing broth was drawn out into a curve so that air was not excluded, but bacteria were trapped in the curve. The flasks were placed in many different places, such as in the middle of Paris, in a country area and on the top of a high mountain in the Alps. Pasteur examined the broth months later for bacteria. He found that the non-living broth had failed to create new life as there were no bacteria in the flasks.

This idea that all living things must arise from other living things was very important to biology and medicine as it led to the work of Joseph Lister.

Joseph Lister (1827–1912)

Lister was an English doctor who recognised the connection between wounds 'going bad' and bacteria. In 1865, Lister began to use disinfectants in his operating theatre in Glasgow. He first used carbolic acid and discovered that wounds did not turn septic after surgery if all the instruments and the hands and clothes of medical staff were disinfected. Lister's work is sometimes regarded as the foundation of modern surgery.

Each step in the development of biological knowledge has depended on the work of previous biologists and has formed the foundation of future work in the sciences.

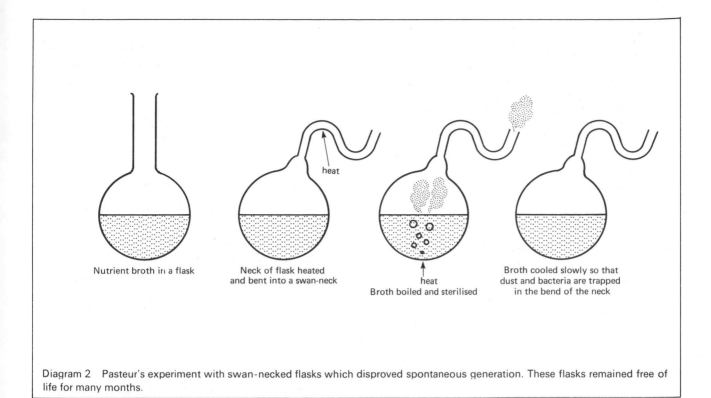

Diagram 2   Pasteur's experiment with swan-necked flasks which disproved spontaneous generation. These flasks remained free of life for many months.

Nutrient broth in a flask	Neck of flask heated and bent into a swan-neck	heat Broth boiled and sterilised	Broth cooled slowly so that dust and bacteria are trapped in the bend of the neck

A carbolic spray being used in an operation in 1882

Related reading

Chapter 19, Circulation of blood
Chapter 65, Agents of disease

Questions I

The answers to questions 1 to 7 are shown by one of the letters A, B, C, D or E.

1   A domestic refrigerator preserves fresh meat for a few days by:

A   killing all the bacteria present.
B   preventing an increase in the numbers of bacteria present.
C   increasing the number of bacteria present.
D   coating the surface of the meat with ice.
E   keeping the food in the dark.

2   Inoculation by loop is used in:

A   immunisation.
B   bacterial culture.
C   blood tests.
D   contraception.
E   sterilisation.

3   A saprophyte is:

A   an animal which lives inside another animal.
B   a plant which lives on another plant.
C   an animal which feeds only on plants.
D   a plant which feeds only on dead organic material.
E   an animal which feeds only on animals.

4   In a parasitic association between two organisms:

A   both members benefit.
B   neither member is harmed.
C   both members suffer harm.
D   only one member benefits.
E   neither member benefits.

5   The antibiotic effect of pencillin was first discovered by:

A   Fleming.
B   Ross.
C   Lister.
D   Simpson.
E   Pasteur.

6   Which of these substances may be used to indicate the presence of carbon dioxide gas?

A   Sodium hydroxide solution
B   Soda lime
C   Potassium hydroxide solution
D   Dilute hydrochloric acid
E   Clear lime water

7   Louis Pasteur is famous for:

A   showing that bacteria exist in the air.
B   discovering penicillin.
C   his theory of evolution.
D   improving the design of microscopes.
E   carrying out experiments on breeding.

8   Copy out and then complete the following sentences:
(i)   That blood circulates around the body was first discovered by ....
(ii)   An example of a saprophyte is ....
(iii)   Rod-shaped bacteria are called ....
(iv)   A parasite is an organism which obtains food from another organism called a ....
(v)   Preserving food by removing water is called ....
(vi)   An external parasite which lives in body hair is called ....
(vii)   An intelligent scientific guess at an answer to a question can be called a ....

9   Suppose that you were interested in the behaviour of a species of snail which is only active at night after rain. It has been suggested that this species only eats lettuces and rarely, if ever, attacks cabbage leaves.
(i)   Describe, in outline only, the methods you would use to discover whether the snails prefer lettuce to cabbage leaves.
(ii)   What observations or measurements would you make during this investigation?
(iii)   What results would you expect if the snails DO prefer lettuce to cabbage?
(iv)   What results would you expect if the snails DO NOT prefer lettuce to cabbage?

(EMREB)

10   Write about six lines on FIVE of the following:
(i)   Refrigeration
(ii)   Antibiotics
(iii)   Care of the skin
(iv)   Dodder
(v)   Saprophytes
(vi)   The three stages of biological investigation
(vii)   William Harvey's hypothesis about the flow of blood

(viii) Types of bacteria
(ix) Canning of food
(x) Karl Linnaeus

11 Write an account about twenty-five lines long on ONE of the following:
(i) Methods of preserving food
(ii) Culturing bacteria
(iii) The life cycle of the pork tapeworm
(iv) The need for controls in biological investigations
(v) Pasteur and his work on spontaneous generation

12 (i) (a) Which two groups of organisms are usually involved in the decay of food?
(b) Suggest three ways in which these organisms could be introduced into the food.
(ii) (a) Describe three ways of killing decay organisms present in food.
(b) Describe two ways in which the activity of decay organisms in food is slowed down.
(iii) (a) State two ways, used in the home, to prevent decay organisms reaching food for short periods of time.
(b) Give two ways to stop decay organisms reaching foods for a long period of time.

(EMREB)

13 (i) (a) Name one animal parasite.
(b) Name the primary host of this parasite.
(c) Describe how the parasite enters and attaches itself to the host, obtains food and reproduces.
(ii) (a) Name a plant parasite.
(b) Name the host of this parasite.
(c) Describe how this parasite obtains its foods.
(iii) (a) What is meant by the term symbiosis?
(b) Give an example of symbiosis.

14 Write brief notes to explain the importance of the work for which each of the following is famous:
(i) Charles Darwin
(ii) Gregor Mendel
(iii) Louis Pasteur
(iv) Joseph Lister
(v) William Harvey

15 Two agar plates were taken and a trench of agar was removed from each plate as shown in the diagram.

Each plate was then streaked with six different types of bacteria (A–F).

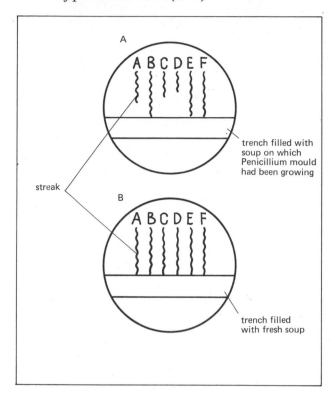

The trench in agar plate A was filled with some soup on which the mould *Penicillium* had been growing.

The trench in agar plate B was filled with fresh soup which had not been in contact with the mould *Penicillium*.

The diagram shows the results after the plates had been placed in a warm place for twenty-four hours.
(i) What can be concluded from the results of this experiment?
(ii) What substance in the soup is likely to have caused the results shown on plate A?
(iii) What name is given to this type of substance?

(EAEB)

Extra reading and book list

Extra reading is provided for pupils who wish to study a topic more fully than some chapters in this book allow. The extra reading listed below for each chapter gives an indication of the areas of study which are needed for this greater depth. In certain chapters this wider understanding can be found within the book; this is indicated by 'See Related Reading'.

A **book list** is also included. Books suggested here are of a standard suitable for extra reading and can be found in most libraries.

Chapter 1   See Related Reading.
Chapter 2   The main characteristics of each plant phyla.
Chapter 3   More detailed characteristics of the animal phyla.
Chapter 4   See Related Reading.
Chapter 5   (i)   Practical details for the food tests.
             (ii)   Some other food tests.
Chapter 6   Variation in the rate of photosynthesis with variation in the level of carbon dioxide and light.
Chapter 7   (i) Different leaf structures and their effect on transpiration rates.
             (ii)   The action of the guard cells in controlling transpiration rate.
             (iii)   Experimental details for measurement of water uptake.
Chapter 8   Further details about saprophytes.
Chapter 9   The chemical structures of carbohydrates, fats and proteins.
Chapter 10   The mouth parts of insects.
Chapter 11   The alimentary canal of some other animals, for example, the rabbit, an insect, an earthworm and a parasite.
Chapter 12   (i)   Experiments to demonstrate protein digestion.
             (ii)   The action of bile.
Chapter 13   (i) The functions of the liver.
             (ii) Perennation in plants.
Chapter 14   Experiments to demonstrate osmosis using parchment and visking tubing.
Chapter 15   (i)   The role of ATP in cell respiration.
             (ii)   Fermentation experiments using yeast.
Chapter 16   See Related Reading.
Chapter 17   Experiments to compare exhaled and inhaled air.
Chapter 18   (i)   Further details of the function of red cells.
             (ii)   Further details of the function of white cells.

Chapter 19   More details of:
             (i)   the lymphatic system.
             (ii)   the formation of blood clots.
             (iii)   blood grouping and the Rhesus factor.
Chapter 20   See Related Reading.
Chapter 21   (i)   The arrangement of xylem and phloem in various plant tissues.
             (ii)   Lenticels in woody stems.
Chapter 22   See Related Reading.
Chapter 23   See Related Reading.
Chapter 24   See Related Reading.
Chapter 25   Leaf structure in desert plants, for example, cactus.
Chapter 26   The comparison of fluid intake and urine output in humans over a period of time.
Chapter 27   Relationship between amount of sweat produced and the amount of urine produced in man.
Chapter 28   The water cycle.
Chapter 29   The sense organs of insects.
Chapter 30   More information about learned behaviour.
Chapter 31   Comparison of different vertebrate brains.
Chapter 32   See Related Reading.
Chapter 33   Further details of the inner ear.
Chapter 34   (i)   Causes and corrections of long and short sight.
             (ii)   Ray diagrams to show light passing into the eye.
Chapter 35   Symptoms of over-activity and under-activity of the endocrine glands.
Chapter 36   (i)   Experiments on geotropisms using a clinostat.
             (ii)   Man-made plant hormones.
Chapter 37   See Related Reading.
Chapter 38   (i)   Fine structure of involuntary and cardiac muscle.
             (ii)   Functions of sphincter muscles in man.
Chapter 39   (i)   Fine structure of bone.
             (ii)   Skeleton of the birds
Chapter 40   See Related Reading.
Chapter 41   (i)   Further details of mitosis.
             (ii)   Meiosis.
Chapter 42   See Related Reading.
Chapter 43   Sexual reproduction in Amoeba and Hydra.
Chapter 44   Puberty and the secondary sex characteristics of the male.
Chapter 45   (i)   Puberty and the secondary sex characteristics of the female.
             (ii)   The corpus luteum.
Chapter 46   Pregnancy in the rabbits or the rats.
Chapter 47   Details and diagrams of metamorphosis in amphibians.
Chapter 48   More details of alternation of generation.
Chapter 49   (i)   The structure of composite flowers.
             (ii)   The advantages and disadvantages of self-pollination.
Chapter 50   (i)   Variety of structure in pollen grains.
             (ii)   Formation and growth of the pollen tube.
Chapter 51   (i)   Hypogeal and epigeal germination.
             (ii)   Details of some germination experiments.
Chapter 52   See Related Reading.
Chapter 53   (i)   The use of the recessive back-cross.
             (ii)   Incomplete dominance.
             (iii)   Inheritance of sex-linked characteristics, for example, haemophilia.
Chapter 54   (i)   Further evidence to support Darwin's theory of natural selection.

Book list

D. G. Mackean, 'Introduction to biology', 5th ed. (John Murray, 1973).

C. V. Brewer and C. D. Burrow, 'Life: form and function' (Macmillan Education, 1972).

C. V. Brewer and C. D. Burrow, 'Principles of biology' (Macmillan Education, 1978).

B. S. Beckett, 'Biology – a modern introduction' (Oxford University Press, 1976).

E. J. Ewington and D. F. Moore, 'General plant and animal biology' (Routledge and Kegan Paul, 1971).

E. J. Ewington and D. F. Moore, 'Human biology and hygiene' (Routledge and Kegan Paul, 1971).

D. G. Mackean and B. Jones, 'Introduction to human and social biology' (John Murray, 1975).

C. Dodds and J. B. Hurn, 'Practical biology', 2nd ed. (Arnold, 1972).

J. W. Lee and D. Martin, 'A Certificate course in practical biology', Vol. 1 'Experimental investigations' (Mills and Boon, 1970).

J. W. Lee and D. Martin, 'A Certificate course in practical biology', Vol. 2 'Plant and general studies' (Mills and Boon, 1971).

Any particular area of interest can be studied in great depth in the following advanced level textbooks:

M. B. V. Roberts, 'Biology – a functional approach', 2nd ed. (Nelson, 1976).

M. B. V. Roberts, 'Biology – a functional approach'. Student's manual (Nelson, 1974).

A. E. Vines and N. Rees, 'Plant and animal biology', Vol. I and Vol. II, 2nd ed. (Pitman, 1972).

Glossary
and
Index

This glossary explains the words that are printed in SMALL CAPITALS the first time they appear in every chapter. The numbers which follow the explanation show the **chapters** in which the word is used.

ABDOMEN. The belly of man, containing stomach, intestines, kidneys and liver. In insects it is the third part of the body, which has no legs. (3, 26)

ABSORBED. The movement of digested food and water through the walls of the intestine and into the blood. (11, 12)

ADAPTATION (ADAPTED). The features of animals and plants which suit them to the place in which they live. (47, 54, 55, 57, 58, 59, 134).

AERATION. The supply and mixing of air. (59)

AGAR. A substance which comes from seaweeds and sets like a jelly. (36, 59, 65)

ALIMENTARY CANAL. The tube running through an animal from mouth to anus. (11, 12, 26, 35)

AREA. See SURFACE AREA.

BILE. A greenish substance which contains certain salts that causes fats to break up into tiny droplets. (11)

BILE PIGMENTS. Coloured substances in the bile that come from the breakdown of worn-out red blood cells. (26)

BOLUS. A ball of chewed food. (10, 11)

CAPILLARY. A narrow tube which carries blood. The walls are only one cell thick. (13, 17, 19, 26, 27, 69, 70)

CARNIVORE. An animal which eats only meat. (10, 55)

CARTILAGE. A firm but elastic TISSUE usually found close to bones. (11, 17, 39, 40)

CELLULOSE. The main part of plant cell walls. (11)

CHARACTERISTICS. The main appearances or features by which something is recognised. (1, 4, 12, 23, 29, 42, 52, 53, 54)

CHLOROPHYLL. The green colour in plants, which takes in light energy for use in PHOTOSYNTHESIS. (2, 6)

CHLOROPLAST. ORGANELLE in plant cells containing CHLOROPHYLL. (4, 6, 7)

CHROMOSOMES. Thread-like structures, made up of DNA, which occur in the NUCLEUS of all cells. (4, 41, 44, 53, 54)

CILIA. Hair-like STRUCTURES which stick out from certain cells. (3, 22, 37, 45)

CONSUMERS. ORGANISMS in a food chain or web which live by eating other organisms. (3, 8, 55)

CONTROL. A second experiment, identical to the first, apart from the one factor which is being investigated. (6, 16, 51, 68)

CONVOLUTED TUBULE. A coiled or twisted small tube. (26)

CO-ORDINATE. Linking together different parts of the body so that they work well together. (35, 38)

COTYLEDON. A special leaf inside a seed, which is often swollen with stored food. (13, 51)

CROSS-POLLINATION. The transfer of pollen from the anther of one flower to the stigma of a flower on another plant. (49, 52)

CYTOPLASM. The contents of a cell inside the MEMBRANE, except the NUCLEUS. (4, 10, 15, 23, 30, 46)

DAUGHTER CELLS. The two cells which result from the division of a single cell by MITOSIS. (41)

DEAMINATION. The removal of nitrogen from unwanted amino acids in the liver. (13)

DENATURING. The breakdown of proteins by EXCESS heat or certain chemicals. (12)

DENTITION. The number and arrangement of teeth in the mouth. (10)

DETRITUS. The dead and decaying remains of plants and animals. (8)

DICOTYLEDONS. A group of plants whose seeds have two seed leaves. (2)

DIFFUSE (DIFFUSION). The movement of molecules from a place of higher concentration to a place of lower concentration. Read chapter 14. (7, 14, 16, 17, 19, 23, 24, 25, 46)

DNA. Deoxyribonucleic acid – the chemical in chromosomes which controls inheritance. (53, 69)

DUCT. A tube which carries fluids. (11, 26, 35, 44)

EMBRYO. The stage of development between the FERTILISED ZYGOTE and the new independent ORGANISM. (12, 13, 44, 45, 46, 51)

ENVIRONMENT. The conditions in which ORGANISMS live. (22, 29, 36, 37, 43, 45, 54, 55, 56, 57, 58, 59, 60, 61, 62, 63, 69, 134)

ENZYME. Proteins which speed up chemical changes in living things. (10, 11, 15, 51, 64)

EXCESS. More of something than is needed. (12, 13, 23, 24, 28)

FAECES. The semi-solid, undigested material which is removed from the body through the anus. (11, 50, 67)

FERTILISATION. The joining together of a male and a female GAMETE to form a ZYGOTE. (43, 44, 45, 46, 47, 49, 50, 51, 53)

FLAGELLA. Long thread-like STRUCTURES which beat. This can produce movement in single-cell ORGANISMS. (3, 37)

FRUIT. Formed from the ovary or receptacle of a flower and containing seeds. (2, 50)

FUNCTION. The way something works and the job that it does. (26, 27, 39, 41, 69)

GAMETES. Cells involved in sexual reproduction such as sperm, POLLEN grains and ova. (43, 44, 48, 49, 50, 52, 53)

GENE. Part of a CHROMOSOME which controls an inherited CHARACTERISTIC. (4, 53, 54)

GENETIC. Concerned with GENES. (41, 42, 53)

GERMINATION. The beginning of growth of a new individual plant. (37, 50, 51)

GRAM (g). A unit for measuring mass. A gram is 1/1000th of a kilogram (kg). (9, 15)

HABITAT. Part of an ENVIRONMENT with one particular set of conditions. (55, 56, 57, 58)

HERBIVORE. An animal which eats only plants. (10, 11, 55)

HORMONE. A chemical messenger produced in one part of an ORGANISM and transported to another part of that organism where it has an effect. (18, 20, 29, 31, 35, 44, 45, 60)

HOST. An ORGANISM infected by a PARASITE. (8, 65, 67)

INCUBATE. Keeping something at a constant and suitable temperature. (45, 65)

JOULE (J). A unit for measuring energy. A joule is 1/1000th of a kilojoule (kJ). (5, 9)

KEY. A list of statements about the CHARACTERISTICS of a group of ORGANISMS which are set out in a way which helps identification. (1, 55, 56)

KILOGRAM (kg). A unit for measuring mass. A kilogram is 1000 grams (g). (9)

KILOJOULE (kJ). A unit for measuring energy. A kilojoule is 1000 joules (J). 4.2 kilojoules = 1 Calorie. (9, 15)

LACTEAL. A tube, in the centre of a VILLUS, which is part of the lymphatic system. (11, 13)

LARVA. An early stage in the life of some animals which is totally unlike the adult. (47, 63)

LIGAMENTS. Strong bands connecting two bones at a joint. (40)

LUBRICANT. A substance which makes surfaces smooth and slippery to overcome friction. (10)

LYMPH. A fluid which comes from plasma, bathes the cells and is returned to the blood system through tubes called the lymphatic system. (11, 12, 19, 31)

MASS. The amount of a particular material present, measured in GRAMS or KILOGRAMS. (8)

MEIOSIS. Cell division that produces GAMETES and the CHROMOSOME number is halved. (53)

MEMBRANE. A thin film or layer enclosing an animal or plant ORGANELLE, cell or group of cells. (4, 12, 14, 17, 20, 24, 25, 30, 33, 40, 44). (See SEMI-PERMEABLE MEMBRANE.)

METABOLISM. The build-up and breakdown of substances inside living cells. (4, 26, 35)

MINERAL SALTS. (MINERALS). Naturally occurring chemicals some of which are essential for healthy growth and development of living things. (5, 8, 9, 18)

MITOCHONDRIA. Rod-shaped STRUCTURES inside cells, concerned with release of energy by respiration. (4, 15)

MITOSIS. Cell division that produces DAUGHTER CELLS and the CHROMOSOME number is unchanged. (41)

MOLECULE. A group of atoms making the smallest part of a particular substance that can exist. (4, 6, 7, 12, 14)

MONOCOTYLEDONS. A group of plants whose seeds have one seed-leaf. (2)

MUCUS. A slimy, sticky fluid SECRETED by special cells. (11, 12, 16, 22, 32, 66)

NUCLEUS. Part of a cell which contains CHROMOSOMES and which controls how the cell will behave. (4, 18, 30, 41, 42, 44, 46, 53, 69)

OMNIVORE. An animal which eats both plants and meat. (10, 11)

ORGAN. A group of TISSUES working together to do a particular job. (4, 17, 22, 41, 46, 49, 67)

ORGANELLE. A STRUCTURE inside a cell which does a particular job. (4, 15)

ORGANISM. Any living animal or plant. (1, 2, 3, 5, 7, 8, 12, 15, 16, 23, 24, 28, 29, 37, 42, 47, 53, 55, 56, 57, 58, 59, 62, 63, 64, 67, 68, 69)

OSMOREGULATION. The control of movement of water in and out of cells by OSMOSIS. (24)

OSMOSIS. The DIFFUSION of water molecules through a SEMI-PERMEABLE MEMBRANE from a weaker to a stronger solution. Read chapter 14. (14, 21, 23, 24, 25)

PARASITE. An ORGANISM living in or on another organism from which it obtains food. (2, 3, 8, 66, 67)

PERISTALSIS. A wave-like motion which moves the food through the ALIMENTARY CANAL. (11, 38)

PHLOEM. Living, tube-like plant cells which allow the movement of food throughout the plant. (13, 21)

PHOTOSYNTHESIS. The process by which green plants use light energy to make food substances from carbon dioxide and water. (2, 6, 7, 13, 15, 21, 25, 28, 37, 42, 55, 57, 62, 67, 69)

POLLEN. Male GAMETES of flowering plants. (37, 49, 50, 63)

POLLINATION. Transfer of pollen from anthers to stigmas in flowering plants. (37, 49, 50, 52) (See also CROSS-POLLINATION and SELF-POLLINATION.)

POLLUTION (POLLUTED). When substances are present in large enough quantities in the ENVIRONMENT to be harmful to living things. (55, 56, 61)

POTOMETER. The apparatus used to measure the uptake of water by a plant shoot. (7)

PRODUCERS. ORGANISMS at the beginning of a food chain or web that make their own food. (2, 6, 8, 55)

PROTOPLASM. The contents of a cell inside the MEMBRANE, including the NUCLEUS. (See CYTOPLASM.)

PROTRUDE. To project or stick out. (49)

REABSORBED. ABSORBED again. (12) (See ABSORBED.)

RECYCLE. To re-use a substance many times. (28, 62)

SALIVA. A fluid produced in the mouth which contains amylase, MUCUS and various chemicals. (10, 12)

SAPROPHYTE. An ORGANISM that feeds on the dead and decaying remains of other organisms. (2, 8, 67)

SATURATED. Soaked with water and unable to hold any more. (7, 16)

SECRETE. The pushing out of substances such as ENZYMES and MUCUS from cells. (33, 35, 40, 41, 44, 45)

SEDIMENTS. Substances which settle to the bottom of liquids. (54)

SELF-POLLINATION. Transfer of POLLEN from the anther of one flower to the stigma of the same flower, or to the stigma of another flower on the same plant. (49, 52)

SEMI-PERMEABLE MEMBRANE. A MEMBRANE which allows the passage of certain substances only and prevents the passage of others. (4, 14, 24, 25) (See MEMBRANE.)

SOLUTE. A substance which is dissolved in a liquid to form a solution. (14)

SOLVENT. A liquid in which other substances can be dissolved to form a solution. (14)

SPECIES. A group of animals or plants which can successfully breed with each other. (1, 3, 44, 50, 54, 56, 57, 58, 60, 62, 68, 69)

SPHINCTER. A ring of muscle in a tube which can narrow or close that tube. (11, 26)

SPORE. A reproductive cell produced by an ORGANISM during asexual reproduction. (2, 48)

STERILISATION. The destruction of all living things by chemicals, heat or radiation. (59, 65)

STIMULUS. Anything which produces a response in an ORGANISM. (29, 30, 32, 33, 36, 37, 38, 41, 44)

STOMATA. The openings in the surface of a leaf through which oxygen, carbon dioxide and water can pass. (7, 16, 23, 25)

STRUCTURE. An object which is built up of many parts. (2, 4, 69)

SUCROSE. A type of sugar. (13)

SURFACE-AREA. The amount of surface which is available for inter-actions to take place. (11, 17, 31)

TENDONS. Strong bands connecting muscles to bones. (38, 40)

THORAX. The chest of man, containing the lungs and heart. In insects, it is the middle part of the body, which has the six legs attached. (3, 17, 20, 39)

TISSUE. A group of cells working together to do a particular job. (4, 12, 17, 20, 29, 30, 31, 41, 44, 46)

TOXIC. Poisonous. (23, 42)

TRANSPIRATION. The loss of water, by evaporation, from plant cells. (7, 14, 16, 21, 25)

TROPISM. The growth movement of a plant in response to a directional, external STIMULUS. (36, 37)

VACUOLE. A fluid-filled space in the CYTOPLASM of a cell. (4, 24)

VASCULAR BUNDLE. A group of cells, including XYLEM and PHLOEM, which transport water and food through a plant. (21, 67)

VESSEL. A tube or canal which carries fluids such as blood, LYMPH and water. (11, 18, 19, 22, 27, 70)

VILLUS (plural – VILLI). Finger-like projection which increases the internal SURFACE-AREA of the small intestine. (11)

WILTING. The collapse of the leaves and stem of a plant due to the loss of water. (7, 25)

XYLEM. Dead tube-like plant cells which conduct water throughout the plant. (21, 25)

ZONATION. The banding of different SPECIES in different parts of the ENVIRONMENT. (58)

ZONE. A band or region. (26, 58)

ZYGOTE. The cell which results from the joining together of a male GAMETE and female GAMETE and from which a new individual can grow. (43, 44, 46, 53)

The authors and publishers wish to acknowledge the following photograph sources:

Cover   Eric Hosking (an eagle owl, *Bubo bubo*).
Heather Angel   pp. 61, 171.
Health Education Council   p. 57.
Eric Hosking   p. 148.
Michael D. Robson   p. 29.
R.T.H.P.L.   p. 165.
Society for Cultural Relations with USSR   p. 142.
C James Webb   pp. 49, 159.

The authors and publishers wish to thank the following, who have given permission for the use of copyright material:

Aldus Books Limited for two maps from *Conservation* by Joyce Jaffe and *Man's Impact on Nature*;

Longman Group Limited for a diagram from Nuffield Biology, Pupil's O-Level text;

Royal College of Physicians for an extract from a report on smoking and health, London 1962;

A. P. Watt on behalf of Barbara Ward and Rene Dubos for a diagram from *Only One Earth*.

The publishers have made every effort to trace copyright holders, but if they have inadvertently overlooked any they will be pleased to make the necessary arrangements at the first opportunity.